Endocrine Pathology

Ricardo V. Lloyd

Endocrine Pathology

With 221 Illustrations and 12 Color Plates

Springer-Verlag
New York Berlin Heidelberg
London Paris Tokyo Hong Kong

Ricardo V. Lloyd, MD, PhD
Associate Professor of Pathology
Warthin-Weller Professor of Pathology
University of Michigan Medical Center
Ann Arbor, Michigan 48109, USA

Library of Congress Cataloging-in-Publication Data
Lloyd, Ricardo V.
Endocrine pathology/Ricardo V. Lloyd.
p. cm.
ISBN 0-387-97166-1
1. Endocrine glands–Histopathology 2. Endocrine glands–Diseases–Diagnosis. I. Title.
[DNLM: 1. Endocrine Diseases–diagnosis. WK 100 L793e]
RC649.L49 1990
616.4′071–dc20
DNLM/DLC 89-22009

Printed on acid-free paper.

Typeset by Publishers Service, Bozeman, Montana.
Printed and bound by Arcata Graphics/Halliday, West Hanover, Massachusetts.
Printed in the United States of America.

9 8 7 6 5 4 3 2 1

ISBN 0-387-97166-1 Springer-Verlag New York Berlin Heidelberg
ISBN 3-540-97166-1 Springer-Verlag Berlin Heidelberg New York

To my wife Debbie for her continuous enthusiasm
and encouragement;
to Vincent, a wonderful son;
to my parents, who provided the environment, support,
and encouragement for my education;
and to my past, present, and future teachers.

Acknowledgments

I would like to express my deepest gratitude to all of my mentors, who have been a great source of inspiration as I have learned and continue to learn more about endocrinology and endocrine pathology. My deepest appreciation is extended to my many colleagues in pathology at the University of Michigan who provided many microscopic slides and photographs and to clinical colleagues who performed detailed clinical evaluations and medical and surgical treatment of patients with endocrine disorders. The help of Dr. K. Kovacs in providing photographs of unusual pituitary tumors is greatly appreciated.

Completion of this book would not have been possible without the excellent secretarial assistance of Rebecca Lentz and Gioia Ameson, whose efforts are greatly appreciated. The photographic assistance of Craig Biddle and Mark Deming and the assistance of Cindy Lam and Joe Mailloux in electron microscopy are greatly appreciated. The assistance of all of the members of my research laboratory who worked diligently in developing and applying some of the techniques discussed in this book is greatly appreciated, with special thanks to Kristina Fields and Dr. Long Jin.

Finally, the continuous support and encouragement of my wife Debbie, son Vincent, many relatives, and parents, especially my mother, Amy Lloyd, are sincerely appreciated.

Contents

CHAPTER 1
The Endocrine System

Historically, the endocrine system was simply considered to be made up of a few glands that secreted their products directly into the bloodstream. These glands were sometimes referred to as "ductless glands." Today the endocrine system is viewed as a very complex system of cells, tissues, and organs producing many hormones and amines. The endocrine system can be divided into two major categories, the diffuse or dispersed neuroendocrine system (DNS) and the nonneuroendocrine system. Most of the cells and tissues that are part of the endocrine system belong to the DNS, while a small group of cells that produce steroid and thyroid hormones have unique characteristics that set them apart from the neuroendocrine system.

The Diffuse Neuroendocrine System (DNS)

The DNS consists of a variety of peptide-and amine-producing cells and neoplasms that are present in many classical endocrine organs and in other tissues with endocrine functions. The term neuroendocrine is used in part because of the similarity of these cells to neurons with respect to cytochemical characteristics and the presence of peptides and amines. There are many modes of neuroendocrine action via peptides or amines produced by the cells of the DNS (Fig. 1.1). These include the classical endocrine action, such as in the anterior pituitary, in which the peptides go directly into the systemic circulation and have an effect at a distant site. In the neuroendocrine mode of action peptides released from neurons reach the local circulation to exert their effects on other endocrine tissues. For example, the hypothalamic releasing hormones which are produced by neurons and secreted into the median eminence, travel to the anterior pituitary to regulate anterior pituitary hormone secretion. In the paracrine mode of action, such as in the gastrointestinal tract and pancreatic islets, peptides are produced by cells that exert a local effect on other target cells by diffusion through the extracellular space. Peptides and amines may also function in the neurotransmitter mode of action. For example, they may be secreted in the central nervous system (CNS) or by ganglia and released into the synaptic cleft through an axodendritic synapse, or they may terminate presynaptically through an axoaxonal synapse (22). Neoplastic endocrine cells often have an autocrine regulatory action in which they produce specific peptides and growth factors that enhance their own growth. Many peptides that are commonly found outside the central nervous system have also been identified in the brain; these include insulin, glucagon, and bombesin. The exact functions of many of these peptides in the CNS are still unknown.

The feedback mechanism is one of the principal regulatory pathways in many endocrine cells (Fig. 1.2). This mechanism often becomes disturbed during the development of endocrine diseases, which may include hypofunction or hyperfunction of endocrine tissues. Hypofunction is commonly associated with destruction of the endocrine tissue, while hyperfunction is associated with the development of hyperplasias or neoplasias.

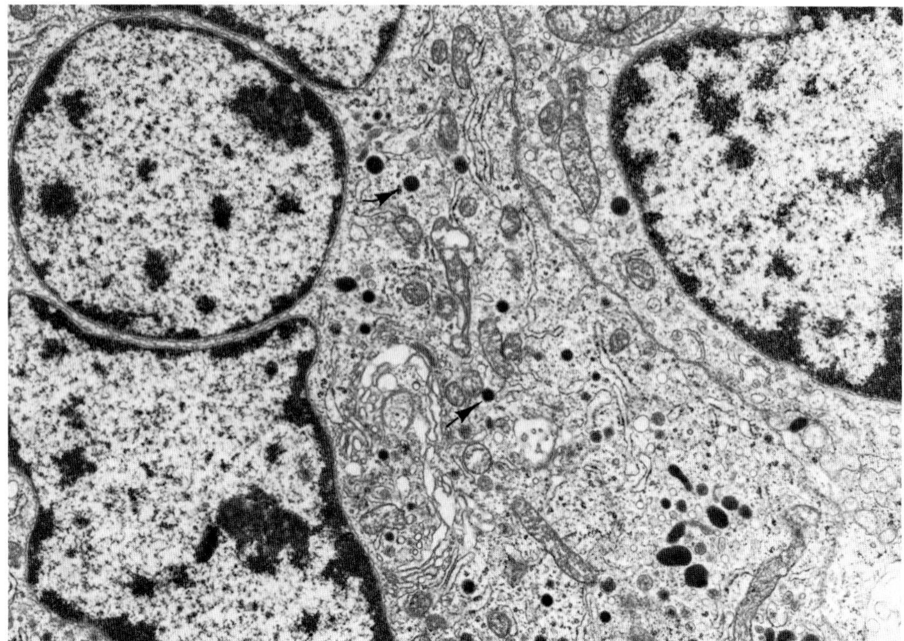

Figure 1.3. Ultrastructural features of a neuroendocrine cell illustrated by a neuroendocrine carcinoma of the lung that produced ACTH and caused Cushing's syndrome. Many secretory granules 100 to 250 nm in diameter (*arrows*) are present in the cytoplasm of the tumor cells (× 8938).

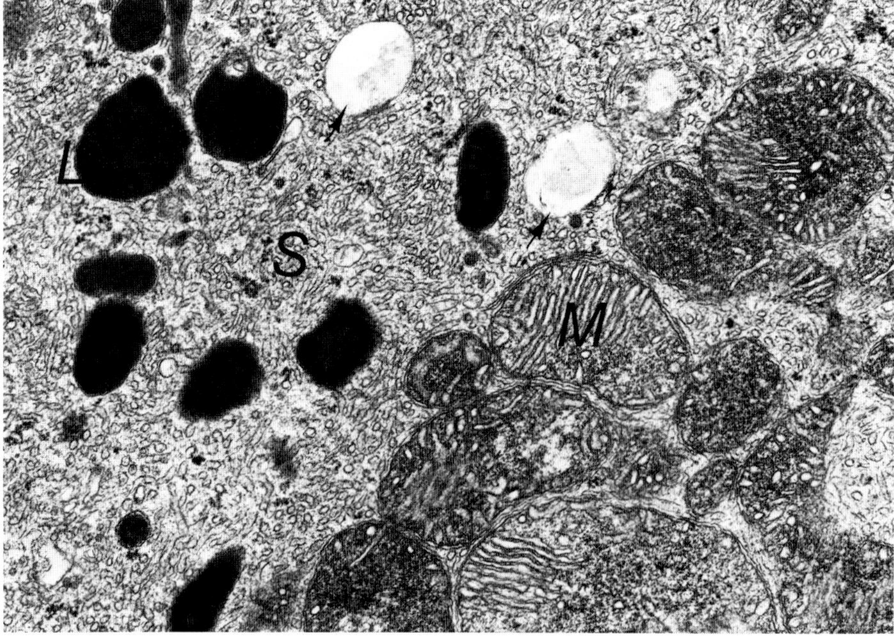

Figure 1.4. Ultrastructural features of a steroid-producing cell illustrated by a cell from the zona fasciculata of the adrenal cortex. Abundant smooth endoplasmic reticulum (*S*) and mitochondria with tubular cristae (*M*) are present. Lipid inclusions (*arrows*) and lysosomes (*L*) are also seen in the cells (× 21,830).

technique is very specific but not very sensitive. In contrast, the argyrophilic reaction is highly sensitive but not very specific. In this silver impregnation technique, endocrine cells can also take up silver salts from ammoniacal silver solutions but cannot utilize the metalic silver without a reducing agent. Commonly used argyrophilic methods include the Grimelius (16) technique and the Bodian, Sevier-Munger, Hellerstrom-Hellman, and Churukian-Schenk methods. The argentaffin reaction is thought to be associated with the presence of serotonin in secretory granules. The binding site of silver in the argyrophilic reaction is not known, although it has been postulated that silver salts may bind to chromogranin proteins (26). Although silver impregnation techniques can help to characterize some cells or neoplasms as neuroendocrine, these methods generally lack the sensitivity and specificity of some of the other methods that will be described.

Electron Microscopy

The identification of secretory vesicles in neuroendocrine cells and their derived neoplasms remains one of the definitive methods of classifying cells and tumors of the DNS. This technique is useful diagnostically (23,24). Ultrastructural studies can readily distinguish between cells and neoplasms that belong to the DNS by detecting secretory granules (Fig. 1.3). Other endocrine cells and tumors can be recognized by the presence of distinct ultrastructural features such as tubular mitochondria in the adrenal cortex and in other steroid-producing cells (Fig. 1.4). The sizes of secretory granules usually range from 100 to 400 nm, although a few cells may have granules over 600 nm in diameter. The morphologic appearance of secretory granules in some normal endocrine cells may be helpful in recognizing the protein or amine products within the granules. Examples include the irregular crystalline arrangement of insulin granules, the dense central core of glucagon granules, and the distinct halo of norepinephrine-containing granules. These morphological features of secretory granules may not be helpful with many neuroendocrine neoplasms, because the ultrastructural features of the neoplasms including the granule morphology are sometimes different from those in the normal cells.

In the diagnosis of some neuroendocrine neoplasms, ultrastructural examination may be more sensitive than some immunohistochemical methods (to be discussed below). For example, when using antibodies that stain secretory granules, a minimum number of secretory granules must be present to detect an immunohistochemical reaction, such as with chromogranin antibodies. On the other hand, the presence of a few secretory granules per cell may be readily detected by electron microscopic examination.

Immunochemistry

The identification and characterization of neuroendocrine tissues have been greatly facilitated by immunochemical staining with specific polyclonal and monoclonal antibodies (Appendix I). The availability of many broad-spectrum antibody markers for cells of the DNS and specific peptides and amines has facilitated the study and diagnosis of neuroendocrine tumors (38). The use of immunochemistry at the electron microscopic level has added another level of specificity to the study of the DNS. With this technique, stored hormone in the individual granules can be readily identified with enzymatic or colloidal gold-linked methods (1). Some of the broad-spectrum markers that are useful in characterizing tissues of the DNS are summarized below.

1. Neuron-specific enolase (NSE). This enzyme, also known as γ-enolase, is a sensitive but not very specific marker for the DNS. NSE is made of two gamma subunits, and it is the most acidic form of the enolase isoenzyme (3,27,40). Some non-neuroendocrine cells and neoplasms react with antisera against NSE. NSE should be used only with other broad-spectrum markers for neuroendocrine cells because of its relatively low specificity. NSE is a cytosolic protein, so the combination of this marker with granule-specific markers such as the chromogranins or 7B2 can add to the characterization of neuroendocrine tumors.

2. Chromogranins. These are acidic proteins that are associated with secretory granules of neuroendocrine cells. The three major chromogranins are designated as chromogranin A, chromogranin B (also known as secretogranin I), and secretogranin II (also known as chromogranin C) (10). Chromogranin A has been studied most extensively in cells

and tumors of the DNS (28,33,44). Chromogranin A is also elevated in the serum of many patients with neuroendocrine neoplasms (33). Because chromogranin antibodies react with the protein in secretory vesicles, neuroendocrine cells or tumors with few secretory vesicles may show a false negative reaction for this marker by immunohistochemistry.

3. Synaptophysin. This 38-kDa antigen is a component of the membrane of presynaptic vesicles. It is widely distributed in neurons and neuroendocrine cells and their neoplasms, including those of the adrenal medulla and pancreas (15). When synaptophysin is used in immunohistochemical characterization of neuroendocrine tumors, the immunoreactivity of certain neoplasms in formalin-fixed, paraffin-embedded sections may be suboptimal, so other fixatives including ethanol should probably be tested (17).

4. Other broad-spectrum neuroendocrine markers. Bombesin is a tetradecapeptide originally isolated from amphibian skin. Gastrin-releasing peptide (GRP), the proposed mammalian analog of bombesin, has been found in many lung and gastrointestinal neuroendocrine cells and tumors (6). LEU-7 (HNK-1) is a monoclonal antibody that was produced against a T-cell leukemia cell line. It reacts with small-cell carcinoma of the lung, pheochromocytoma, and other neuroendocrine neoplasms in addition to lymphoid cells (7,43).

Protein 7B2 was originally extracted from porcine and human pituitary (19). This protein occurs in most types of neuroendocrine cells and neoplasms, with the highest concentrations noted in pancreatic beta cell neoplasms (41). Protein PGP 9.5 is another broad-spectrum marker that has been found in neurons and in many cells of the DNS (42). Monoclonal antibody HISL-19 is another helpful probe for neuroendocrine differentiation (21).

With such a wide spectrum of neuroendocrine markers available, characterization of neuroendocrine cells and tumors with a battery of these antibodies in addition to the use of more specific antibody markers is now done routinely (5).

In Situ Hybridization (ISH) Histochemistry

Localization of the messenger ribonucleic acid (mRNA) for specific peptides and other neuroendocrine markers can be useful in characterizing neuroendocrine cells and neoplasms (9,25) (Appendix II). Detection of the mRNA can be extremely helpful, especially when a neuroendocrine cell or tumor does not contain detectable amounts of protein hormone. ISH can also be used to obtain information about intracellular hormone synthesis and to distinguish between de novo synthesis and uptake, because the presence of the mRNA within the cell usually indicates that the hormone is being synthesized by a particular cell. ISH can be combined with immunochemistry to analyze mRNA and protein expression by the same cells concurrently. It can also be used to study synthesis of ectopic mRNA by certain neoplasms in which the ectopic hormone may not be stored in sufficient quantities to be detected. Ultrastructural studies with ISH can also be done to study the sites of synthesis and processing of specific mRNAs (2).

Flow Cytometry and Cytophotometric DNA Analysis

Many investigators have applied flow cytometric and cytophotometric techniques to analyze the nuclear DNA ploidy of endocrine cells (12,18, 20,31). Many malignant tumors contain an aneuploid or abnormal nuclear DNA content. Flow cytometric studies of pheochromocytomas (18) and of Hürthle cell tumors (31) have revealed a higher percentage of aneuploid peaks in malignant tumors. Similar studies with cytophotometric techniques have been reported (12). However, the usefulness of DNA aneuploidy as a definitive sign of malignancy has been questioned by other workers (20), since in one study many benign tumors including pituitary and parathyroid adenomas contained aneuploid peaks and none of the adenomas gave rise to metastases after conservative surgical treatment (20).

Endocrine Cells and Tumors That Are Not Members of the DNS

The steroid-producing endocrine cells of the adrenal cortex, ovary, and testis are not members of the DNS. Likewise, the thyroid follicular cells that produce thyroxine and related hormones do not form

Table 1.2. Endocrine tissues that are not members of the diffuse neuroendocrine system

Tissues	Hormones in normal cells and neoplasms
Thyroid follicular cells	Thyroxine, triiodothyronine, thyroglobulin
Adrenal cortex	Glucocorticoids, mineralocorticoids, sex steroids
Testis	Androgens, proopiomelanocortin, inhibin, human chorionic gonadotropin
Ovary	Estrogens, androgens, human chorionic gonadotropin, inhibin

the DNS (Table 1.2). The steroid-producing cells have distinct ultrastructural features including tubulovesicular mitochondria and abundant smooth endoplasmic reticulum. Nonneuroendocrine cells do not have dense-core secretory granules and they do not express broad-spectrum neuroendocrine markers. Very few specific immunohistochemical markers are available to characterize steroid-producing cells. Steroid metabolizing enzymes such as 21-hydroxylase enzymes can be found in these cells but are not specific for steroid-producing cells (39). The thyroid follicular cells often contain thyroglobulin, which can be used as a specific immunochemical marker. The importance of separating cells and tumors of the DNS from this group is related to the observation that, in general, diseases affecting the DNS do not involve thyroid follicular cells and steroid-producing cells. Diseases such as the multiple endocrine neoplasia syndromes and ectopic hormone production are usually present mainly in tumors derived from cells that form part of the DNS.

References

1. Bendayan M (1982) Double immunocytochemical labeling applying the protein A-gold technique. J Histochem Cytochem 30:81–85
2. Binder M, Tourmente S, Roth J, Renaud M, Gehring WJ (1986) *In situ* hybridization at the electron microscope level: Localization of transcripts on ultrathin section of Lowicryl K4M-embedded tissue using biotinylated probes and protein A-gold complexes. J Cell Biol 102:1646–1653
3. Bishop AE, Polak JM, Facer P, Ferri GL, Marangos PJ, Pearse AGE (1982) Neuron-specific enolase: A common marker for the endocrine cells and innervation of the gut and pancreas. Gastroenterology 83: 902–915
4. Boland RP (1974) The neurocristopathies: A unifying concept of disease arising in neural crest maldevelopment. Hum Pathol 5:409–429
5. Bordi C, Paolo F, D'Abba T (1988) Comparative study of seven neuroendocrine markers in pancreatic endocrine tumours. Virchows Arch [Pathol Anat] 413:387–398
6. Bostwick DG, Roth KA, Evans DJ, Barchas JD, Bensch KG (1984) Gastrin-releasing peptide, a mammalian analog of bombesin, is present in human neuroendocrine lung tumor. Am J Pathol 117:195–200
7. Bunn PA, Linnoila I, Minna JD, Carney D, Gazdar AF (1985) Small cell lung cancer, endocrine cells of the fetal bronchus, and other neuroendocrine cells express the LEU-7 antigenic determinant present on natural killer-cells. Blood 65:764–768
8. DeLellis RA (1971) Formaldehyde-induced fluorescence technique for the demonstration of biogenic amines in diagnostic histopathology. Cancer 28: 1704–1710
9. DeLellis RA, Wolfe HJ (1987) New techniques in gene product analysis. Arch Pathol Lab Med 111: 620–627
10. Eiden LE, Huttner WB, Mallet J, O'Conner DT, Winkler H, Zanini A (1987) A nomenclature proposal for the chromogranin/secretogranin proteins. Neuroscience 21:1019–1021
11. Feyrter F (1953) Über die peripheren endokrinen (parakrinen) Drüsen des Menschen. Wien-Düsseldorf, Verlag für Medizinische Wissenschaften, Wilhelm Maudrich
12. Flint A, Davenport RD, Lloyd RV, Beckwith AL, Thompson NW (1988) Cytophotometric measurements of Hürthle cell tumors of the thyroid gland. Correlation with pathologic features and clinical behavior. Cancer 61:110–113
13. Fontaine J, Le Douarin NM (1977) Analysis of endoderm formation in the avian blastoderm by the use of quail-chick chimeras. J Embryol Exp Morphol 41: 209–222
14. Fujita T (1977) Concept of paraneurons. Arch Histol Jpn (Suppl) 40:1–12
15. Gould VE, Wiedenmann B, Lee I, Schwecheimer K, Dockhorn-Dworniczak B, Radosevich JA, Moll R, Franke WW (1987) Synaptophysin expression in neuroendocrine neoplasms as determined by immunocytochemistry. Am J Pathol 126:243–257
16. Grimelius L (1968) A silver nitrate stain for α_2 cells in human pancreatic islets. Acta Soc Med Upsal 73:243–270

17. Hoog A, Gould VE, Grimelius L, Franke WW, Falkmen S, Chejfec G (1988) Tissue fixation methods after the immunohistochemical demonstrability of synaptophysin. Ultrastruct Pathol 12:673–678
18. Hosaka Y, Rainwater LM, Grant CS, Farrow GM, Van Heerden JA, Leiber MM (1986) Pheochromocytoma: Nuclear deoxyribonucleic acid patterns studied by flow cytometry. Surgery 100:1003–1010
19. Hsi KL, Seidah NG, Deserres G, Chretien M (1982) Isolation and NH_2-terminal sequence of a novel porcine anterior pituitary polypeptide. Homology to proinsulin, secretin and Rous sarcoma virus transforming protein TVFV60. FEBS Lett 147:261–266
20. Joensum H, Klemi PJ (1988) DNA aneuploidy in adenomas of endocrine organs. Am J Pathol 132: 145–151
21. Kirsch K, Bubbaum P, Horvat G, Krisch K, Neuhold N, Ultrich W, Srikanta S (1986) Monoclonal antibody HISL-19 as an immunocytochemical probe for neuroendocrine differentiation. Its application in diagnostic pathology. Am J Pathol 123:100–108
22. Krieger DT, Martin JB (1981) Brain peptides. N Engl J Med 304:876–885
23. Lechago J (1982) The endocrine cells of the digestive and respiratory systems and their pathology. In: Bloodworth JMB Jr (ed) Endocrine Pathology, General and Surgical, 2nd ed. Baltimore, Williams & Wilkins, pp 513–555
24. Le Douarin NM (1982) The Neural Crest. Cambridge, England, Cambridge University Press
25. Lloyd RV (1987) Use of molecular probes in the study of endocrine diseases. Hum Pathol 18:1199–1211
26. Lloyd RV, Mervak T, Schmidt K, Warner TFCS, Wilson BS (1984) Immunohistochemical detection of chromogranins and neuron-specific enolase in pancreatic endocrine neoplasms. Am J Pathol 8:607–614
27. Lloyd RV, Warner TF (1984) Immunohistochemistry of neuron-specific enolase. In: DeLillis RA (ed) Advances in Immunochemistry. New York, Masson, pp 127–140
28. Lloyd RV, Wilson BS (1983) Specific endocrine tissue marker defined by a monoclonal antibody. Science 222:628–630
29. Mackay B, Osborne BM (1978) The contributions of electron microscopy to the diagnosis of tumors. Pathol Annu 8:359–405
30. Masson P (1914) La glande endocrine de l'intestin chez l'homme. CR Acad Sci 158:52–61
31. McLeod MK, Thompson NW, Hudson JL, Gaglio JA, Lloyd RV, Harness JK, Nishiyama R, Cheung PSY (1988) Flow cytometric measurements of nuclear DNA and ploidy analysis in Hürthle cell neoplasms of the thyroid. Arch Surg 123:849–854
32. Obendorfer S (1907) Karzinoide Tumoren des Dünndarms. Frankf Z Pathol 1:426–432
33. O'Connor DT, Deftos LJ (1986) Secretion of chromogranin A by peptide-producing endocrine neoplasms. N Engl J Med 314:1145–1151
34. Pearse AGE (1966) Common cytochemical properties of cells producing polypeptide hormones with particular reference to calcitonin and the thyroid C cells. Vet Rec 79:587–590
35. Pearse, AGE, Takor-Takor T (1974) Embryology of the diffuse neuroendocrine system and its relationship to the common peptides. Fed Proc 38:27–41
36. Pearse AGE (1974) The APUD cell concept and its implication in pathology. Pathol Annu 9:27–41
37. Pearse AGE (1986) The diffuse neuroendocrine system: Peptides, amines placodes and the APUD Theory. Prog Brain Res 68:25–31
38. Polak JM, Van Noorden S (eds) (1986) Immunocytochemistry: Modern Methods and Applications, 2nd ed. Bristol, John Wright and Sons
39. Sasano H, Ohkubo T, Sasano N (1988) Immunohistochemical demonstration of steroid C-21 hydroxylase in normal and neoplastic salivary glands. Cancer 61:750–753
40. Schmechel P, Marangos PJ, Brightman M (1978) Neuron-specific enolase is a molecular marker for peripheral and central neuroendocrine cells. Nature 276:834–836
41. Suzuki H, Ghatei MA, Williams SJ, Uttenthal LO, Facer P, Bishop AE, Polak JM, Bloom SR (1986) Production of pituitary protein 7B2 immunoreactivity by endocrine tumors and its possible diagnostic value. J Clin Endocrinol Metab 63:758–765
42. Thompson RJ, Doran JF, Jackson P, Dhillon AP, Rhodes J (1983) PGP 9.5—a new marker for vertebrate neurons and neuroendocrine cells. Brain Res 278:224–228
43. Tischler AS, Mobtaker H, Mann K, Nunnemacher G, Jason WJ, Dayal Y, Delellis RA, Adelman L, Wolfe HJ (1986) Anti-lymphocyte antibody LEU-7 (HNK-1) recognizes a constituent of neuroendocrine granule matrix. J Histochem Cytochem 34: 1213–1216
44. Wilson BS, Lloyd RV (1984) Detection of chromogranin in neuroendocrine cells with a monoclonal antibody. Am J Pathol 115:458–468

CHAPTER 2
Pituitary Gland and Hypothalamus

Development and Anatomy

The human pituitary gland has two major parts—the adenohypophysis, which comprises the pars distalis, pars intermedia, and pars tuberalis, and the neurohypophysis, which includes the posterior lobe, neural stalk, and infundibulum. The adenohypophysis is derived embryologically from the invagination of the stomodeal ectoderm, Rathke's pouch, which is the roof of the stomodeum or primitive buccal cavity. The neurohypophysis originates in the floor of the diencephalon. It retains a permanent connection with the brain via the infundibular stalk. The pituitary can be recognized macroscopically by 12 weeks of gestation (30).

Pearse considered the adenohypophysis to be of neuroectodermal origin, derived from the neural ridge placodes or local thickenings of the general head and trunk ectoderm of the embryo (50). Although all of the cells of the adenohypophysis have been included in the DNS and share many cytochemical and ultrastructural features with other neuroendocrine cells, only the corticotrophs have been shown to have amine-handling functions.

About the third month of uterine life, acidophilic and basophilic cells can be identified in the pituitary (19,30). All of the cell types present in the adult adenohypophysis can be identified immunohistochemically in the fetal pituitary (6). The pituitary gland is a bilaterally symmetrical tan-red gland. It is located in the sella turcica, which is a midline cavity of the sphenoid bone. The pituitary weighs approximately 100 mg at birth, while the adult gland weighs between 400 and 900 mg. The anterior lobe makes up 80% of the pituitary (Fig. 2.1). The weight of the pituitary gland increases during pregnancy secondary to prolactin cell hyperplasia. The gland may weigh more than 1000 mg during late pregnancy. The pituitary gland of multiparous women usually weighs more than the gland of nulliparous women.

Extrasellar adenohypophyseal tissue may be present along the path of fetal development of the pars distalis. Ectopic or pharyngeal pituitary can be found at autopsy in most cases (10,45) and may occasionally give rise to adenomas.

The pituitary is covered by the dura mater, part of which is the diaphragma sellae. An incomplete or absent sellar diaphragm may lead to the "empty sella" syndrome, in which the pituitary is compressed secondary to cerebrospinal fluid pressure because of the opened sellar diaphragm (29).

The blood supply to the pituitary is from branches of the internal carotid arteries. These include the superior hypophyseal arteries, which give rise to capillaries, and the gomitoli, which are abundant in the infundibulum and hypophyseal stalk (30). The long portal vessels and the short vessels carry about 70 to 90% of the blood to the adenohypophysis. The middle and inferior hypophyseal arteries are connected to the neurohypophysis. The hypothalamic hormones are transported to the median eminence along nerve fibers and end in the perigomitolar capillaries and in the neural lobe.

The posterior pituitary has contact with the hypothalamus. It is composed of nerve fibers, pituicytes, and axon terminals with neurosecretory material containing mostly vasopressin and oxytocin and the carrier protein neurophysin.

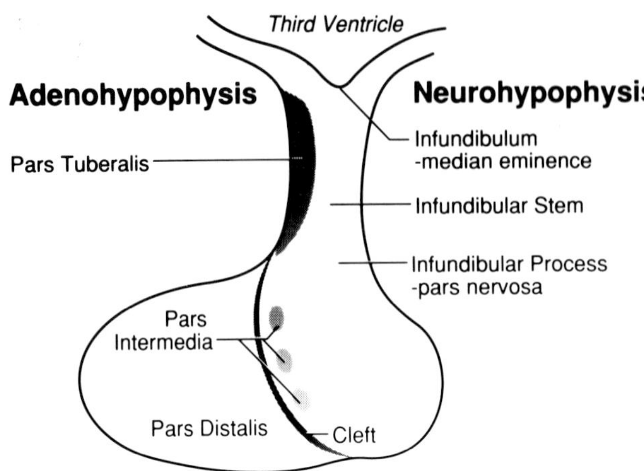

Figure 2.1. Diagram showing relationship of the adenohypophysis and neurohypophysis.

The adenohypophysis has minimal innervation with a few sympathetic nerve fibers that penetrate the anterior lobe adjacent to capillaries. The neural connections to the posterior pituitary are very important, since severe atrophy of the neurohypophysis occurs after interruption of hypothalamic innervation (15).

The Normal Gland

Hypothalamic hormones play a critical role in regulating anterior pituitary secretion and synthesis. These peptides and amines include corticotropin releasing hormone (CRH), gonadotropin releasing hormone (GnRH), thyrotropin releasing hormone (TRH), growth hormone releasing hormone (GHRH), somatostatin (SRIF), vasoactive intestinal polypeptide (VIP), and dopamine (Fig. 2.2). The main features of these hormones are summarized in Table 2.1.

Growth hormone (GH) cells are the most abundant cell types in the adenohypophysis and constitute approximately 50% of the total cells (Fig. 2.3; Table 2.2). In general, the GH cells are located in the lateral wings of the anterior lobes. Immunochemical staining for GH is usually intense in these medium-sized cells. Electron microscopic studies reveal secretory granules 250 to 700 mn in diameter (Fig. 2.4). The number, distribution, and morphology of the GH cells are fairly consistent and are not greatly affected by age, sex, or disease.

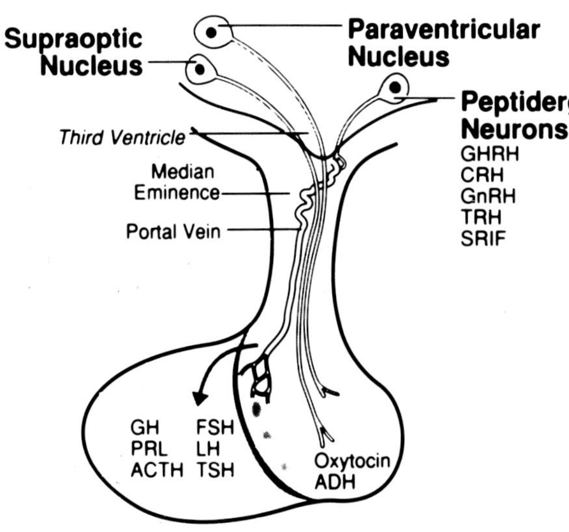

Figure 2.2. Diagram showing the peptidergic neurons with releasing hormones that regulate anterior pituitary hormone secretion and the neurons that produce oxytocin and vasopressin or antidiuretic hormone (ADH), which are stored in and released from the posterior pituitary.

Prolactin (PRL) cells are the second most common cell type. They constitute 15 to 25% of anterior pituitary cells and are distributed throughout the pars distalis (Fig. 2.5). These cells are usually acidophilic or chromophobic on hematoxylin and eosin (H & E) stains. PRL cells are numerous in the fetus and neonate, possibly due to maternal estrogens. The number of PRL cells decreases after birth and remains low during childhood. The highest percentage of PRL cells is found during pregnancy and lactation secondary to hyperplasia (Fig. 2.6). The weight of the pituitary gland may be more than 1000 mg during this period. Immunochemical staining and electron microscopy of the normal pituitary show two types of PRL cells. One type stains strongly and diffusely for PRL; the second type stains in a juxtanuclear immunoreactive staining pattern. Ultrastructurally, the first cell type has large secretory granules 500 to 700 nm in diameter. The second type has secretory granules 200 to 350 nm in diameter. A mixed PRL-GH cell or mammosomatotroph has been identified in the normal human pituitary by the reverse hemolytic plaque assay (39). This cell type is also present in other vertebrates (49).

Corticotroph cells constitute about 15% of anterior pituitary cells. These cells are periodic acid-Schiff (PAS) positive and are located mainly

Table 2.1. Hypothalamic hormones

Hormone[a]	Amino acids	Principal function
CRH	41	Release of ACTH
GnRH	10	Release of LH and FSH
GHRH	40	Release of GH
TRH	3	Release TSH and PRL
SRIF	14	Inhibition of GH release
VIP	28	Putative PRL releasing hormone
Dopamine	–[b]	Inhibition of PRL release

[a]Abbreviations: CRH, corticotropin releasing hormone; GnRH, gonadotropin releasing hormone; GHRH, growth hormone releasing hormone; TRH, thyrotropin releasing hormone; SRIF, somatostatin; VIP, vasoactive intestinal polypeptide.
[b]Amine rather than a peptide.

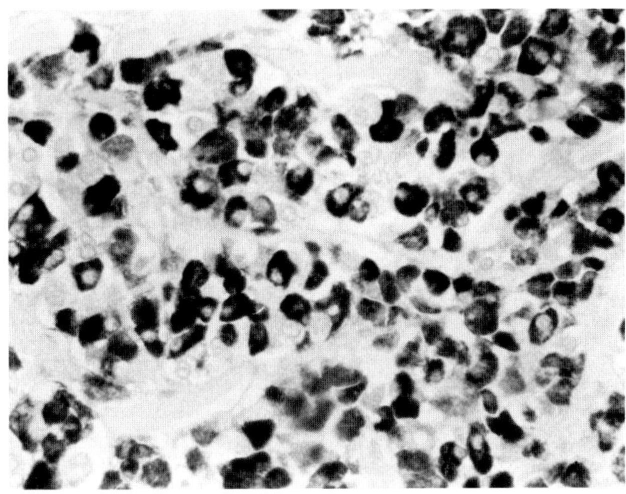

Figure 2.3. Normal pituitary showing GH cells, which constitute about 50% of anterior pituitary cells.

Table 2.2. Anterior pituitary cells

Cell type	Molecular weight	Percent of cells	Secretory granule diameter (nm)
Growth hormone	21,800	50	250–600
Prolactin	22,500	15–25	500–700
Adrenocorticotropin	4,500	15–20	200–350
Follicle-stimulating hormone	29,000	10[a]	200–300
Luteinizing hormone	29,000	10[a]	300–500
Thyroid-stimulating hormone	29,000	5	100–300

[a]Most gonadotropic cells contain both FSH and LH.

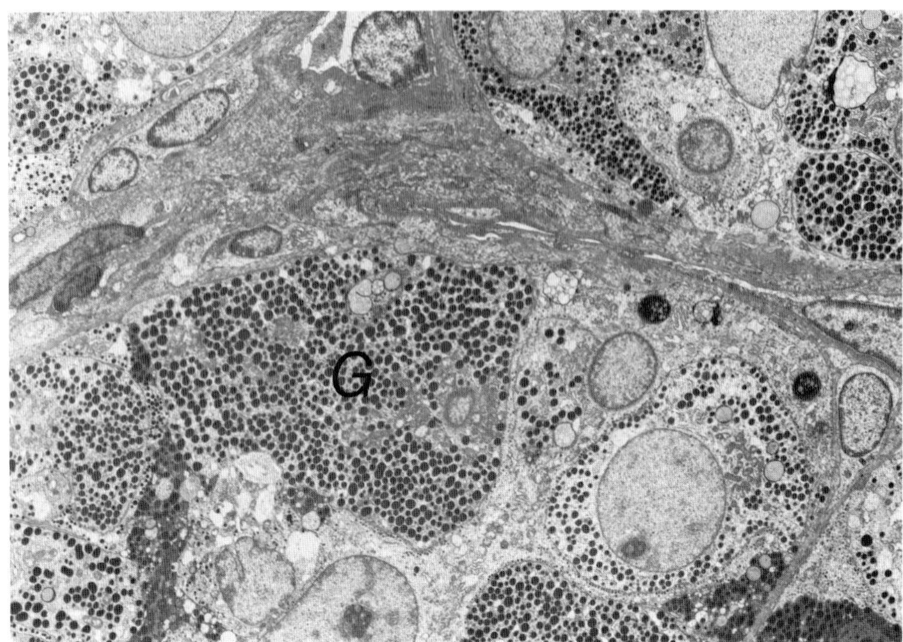

Figure 2.4. Electron micrograph of normal pituitary showing several GH (*G*) and other cell types (×2100).

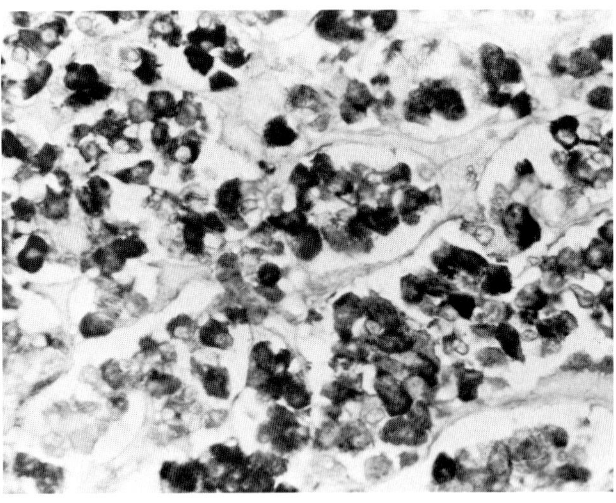

Figure 2.5. PRL cells in normal pituitary. These cells make up about 15 to 25% of total cells.

in the central mucoid wedge. Immunochemistry reveals strong staining for ACTH, β-lipotropin, endorphin, and α- and β-melanocyte-stimulating hormone (MSH) (Fig. 2.7). The "enigmatic body" or large perinuclear vacuole present in corticotrophs probably represents lysosomes and does not stain for ACTH. Ultrastructural studies reveal secretory granules 250 to 700 nm in diameter and type I microfilaments within the cytoplasm adjacent to the nucleus. Crooke's hyaline change (14,33), represented by a glassy homogeneous material in the cytoplasm that pushes the ACTH immunoreactive granules to the periphery, is often associated with glucocorticoid excess. This

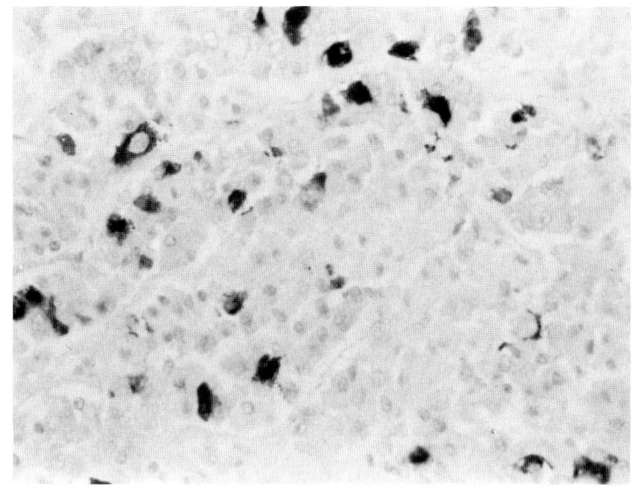

Figure 2.6. Pituitary from 4-month pregnant woman showing PRL cell hyperplasia.

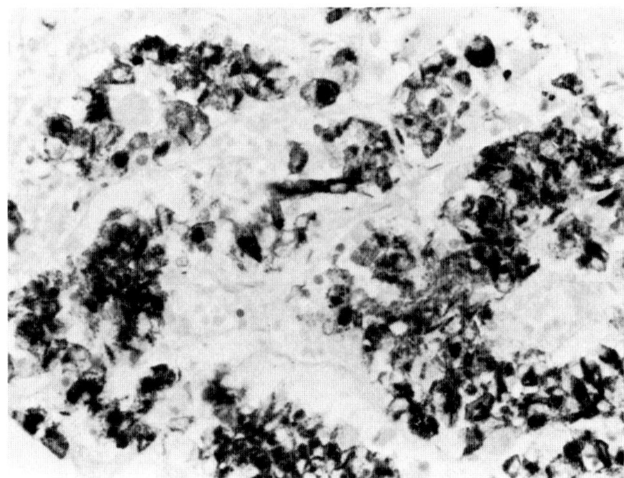

Figure 2.7. Corticotropic cells in normal pituitary. These cells make up 15 to 20% of the pituitary cells.

material is composed of type I microfilaments and stains positively for cytokeratin intermediate filament proteins (48).

Gonadotrophs make up approximately 10% of anterior pituitary cells and are often PAS positive. Immunochemical staining often reveals both FSH and LH in the same cells, although a small percentage of cells contain only FSH or LH (51) (Fig. 2.8). Ultrastructural studies reveal two granule populations, 200 to 300 nm and 300 to 500 nm in diameter. Long-standing gonadectomy results in "signet-ring" or "castration" cells which contain widely dilated rough endoplasmic reticulum (RER) and decreased numbers of secretory granules.

Thyrotrophs constitute 5% of anterior pituitary cells. They are also PAS positive. Immunostaining shows many angular cells with cytoplasmic TSH (Fig. 2.9). Ultrastructural studies show small secretory granules 100 to 300 nm in diameter, which are usually adjacent to the cell membranes. Chronic hypothyroidism often leads to "thyroidectomy cells," consisting of enlarged thyrotrophs with prominent Golgi complexes, dilated RER, and decreased numbers of secretory granules.

Follicular cells or folliculostellate cells are nonsecretory cells that are present in the anterior pituitary. These cells are agranular and contain prominent junctional complexes and a central

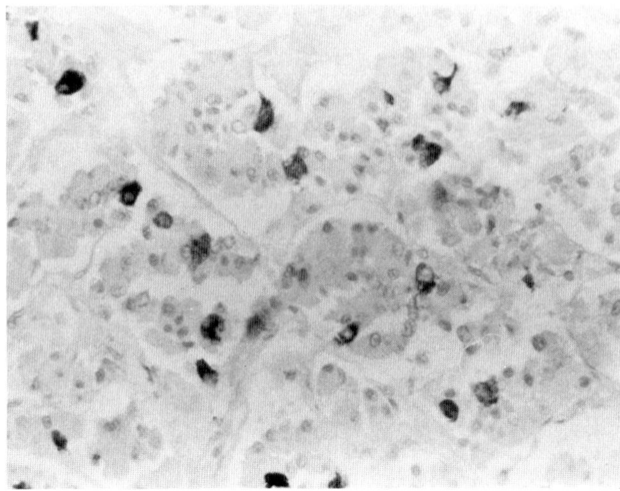

Figure 2.8. Gonadotropic cells stained for FSH. Both FSH and LH are located in the same cells, which constitute about 10% of anterior pituitary cells.

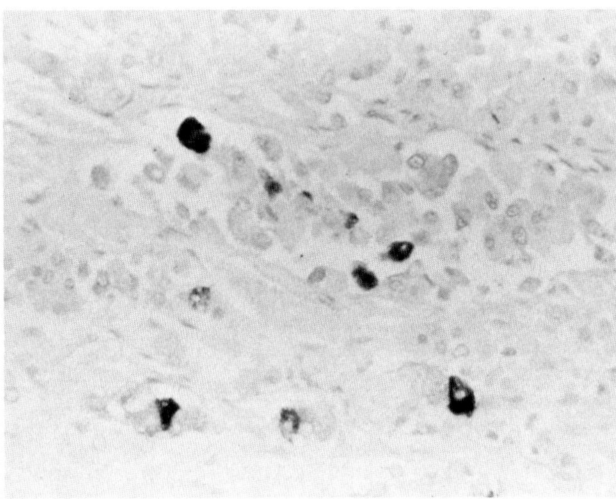

Figure 2.9. Angular TSH cells in the anterior pituitary. These constitute about 5% of anterior pituitary cells.

lumen. Immunostaining reveals S100 protein immunoreactivity (47) (Fig. 2.10). These cells contain fibroblast growth factor (16) and have paracrine functions (5) as well as phagocytic functions (72) in some species.

The pars distalis, which forms part of the adenohypophysis, is covered by arachnoid membranes and contains immunoreactive FSH and LH cells with occasional ACTH or TSH cells (2). The pars intermedia also forms part of the adenohypophysis. It is not well developed in humans and consists of cavities of varying sizes lined by a single layer of cuboidal partly ciliated epithelium (30). The lumens of these cavities are filled with colloidal material that shows ACTH immunoreactivity. Basophil cells that extend into pars nervosa and deeply into the neural lobe are often seen in pituitary tissues from older individuals, a condition termed "basophil invasion." These cells are often positive for ACTH.

The posterior lobe of the pituitary is part of the nervous system. It contains neurosecretory granules with oxytocin, vasopressin, neurophysin carrier protein, and pituicytes that have a supporting function. The posterior pituitary hormones are produced in the supraoptic and paraventricular

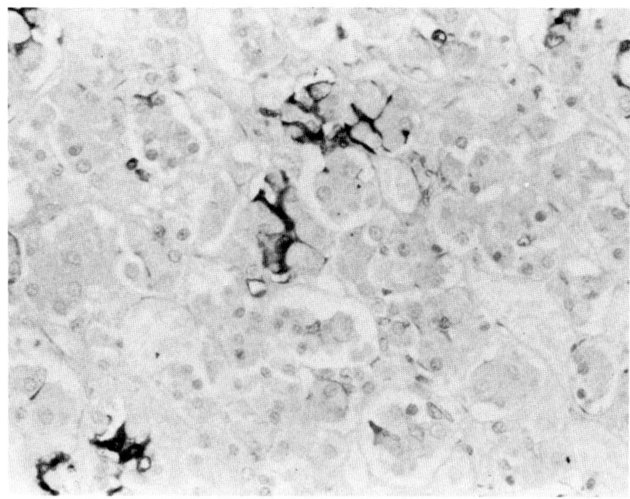

Figure 2.10. Folliculostellate cells with cytoplasmic projections extending between secretory cells.

nuclei and then transported to the posterior lobe for storage and subsequent release. "Herring bodies" or dilated nerve endings with stored hormones can be seen on ultrastructural examination.

Pathology of the Adenohypophysis

Hypoplasia

Congenital hypoplasia of the pituitary gland has been described by several authors (58). In anencephalic infants the pituitary gland is small. Corticotroph and GH cells can be identified by immunostaining (8).

Hemorrhage

Intracranial hemorrhage, usually secondary to cerebrovascular accidents and trauma, may extend into the subarachnoid space of the pituitary.

Infarction

Small infarcts in the adenohypophysis may be seen incidentally at autopsy in many disorders. The most common conditions include severe atherosclerosis. Prolonged use of respirators for assisted ventilation may lead to infarction. Infarcts may also be seen in diabetes mellitus, with tumors, and with inflammatory diseases. Sheehan's syndrome is a specific type of infarction. In this disorder there is severe hypotension, most often associated with hemorrhage and shock at the time of delivery. This leads to severe thrombosis in the sinusoids of the pituitary and infarction of the adenohypophysis (68). The posterior lobe is often preserved. The severity of hypopituitarism depends on the extent of destruction. Usually 80 to 90% of the adenohypophysis must be destroyed before the patient is symptomatic. Failure of lactation may be noted first, followed by adrenal cortical deficiency, amenorrhea, and hypothyroidism.

Simmonds' Disease

This is a historical term also referred to as "pituitary cachexia." It is the clinical syndrome of hypopituitarism in adults and may have many etiologies, including infarction and metastatic malignancies.

Granulomas

Granulomatous inflammation involving the pituitary may be caused by many conditions including tuberculosis, sarcoidosis, and histiocytosis X.

Autoimmune Hypophysitis

Autoimmune hypophysitis is associated with an infiltrate of mononuclear cells, chiefly lymphocytes in the pituitary. This condition is usually part of a spectrum of autoimmune diseases involving the pituitary (1,44).

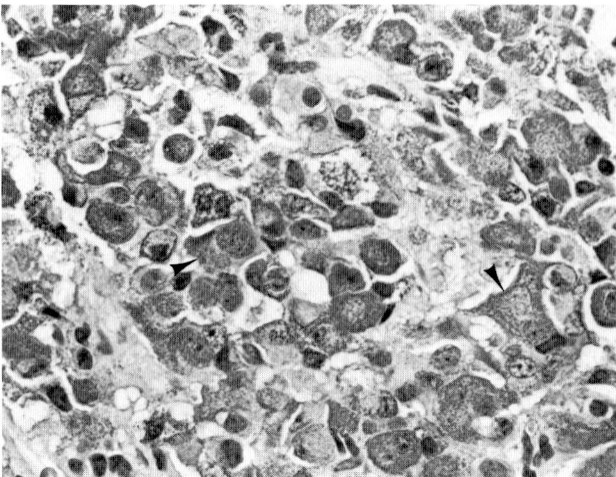

Figure 2.11. GH cell hyperplasia from the pituitary gland of a patient with a pancreatic tumor that was secreting growth hormone releasing hormone. The GH cells have hypertrophied cytoplasm and large nuclei (*arrows*).

Pituitary Hyperplasia

Hyperplasia of one pituitary cell type without a neoplasm leading to specific clinical signs or symptoms is very uncommon (Fig. 2.11). Hyperplasia may occasionally be seen at autopsy as an incidental finding. Hyperplasia is extremely difficult to diagnose in small surgically excised specimens. The distribution of the various adenohypophyseal cell types must be considered when analyzing a pituitary biopsy for evidence of hyperplasia. Hyperplasia can be diffuse or nodular. A recent study of 18 pituitaries from patients with untreated Addison's disease revealed both diffuse and nodular ACTH cell hyperplasia (64). Untreated hypothyroidism can also lead to TSH and PRL cell hyperplasia (53,65). Cases of idiopathic hyperplasia of ACTH cells have also been reported (43).

The reticulin stain can be used to distinguish between a small adenoma and hyperplasia. In general, unlike an adenoma, hyperplasia is not associated with a pseudocapsule. The acinar pattern is lost in adenomas but is retained in hyperplasia. Most cases of diffuse hyperplasia causing a specific clinical disorder can be diagnosed with certainty only if the entire pituitary is available for examination. However, nodular hyperplasia secondary to hypothalamic hormone stimulation has been well documented in biopsied pituitaries (42). In such cases enlarged nodules of pituitary cells with preservation of the acinar architecture are often seen. Reticulin stains in cases of nodular hyperplasia often show the delicate reticulin network of the normal pars distalis, although the pattern is distorted.

Pituitary Adenomas

Introduction

Pituitary neoplasms are usually benign. The etiology of these neoplasms in humans is unknown. Although in animal models excessive estrogens can lead to the development of PRL cell hyperplasia and PRL-producing neoplasms, evidence that estrogens can cause pituitary adenomas in humans is mostly anecdotal (20). Estrogens do stimulate growth of existing prolactinomas (76). Disruption of the normal feedback mechanisms by ablation of the target organ may be associated with hyperplasia and pituitary tumor development such as in long-standing hypothyroidism and Addison's disease (64).

Incidental pituitary tumors are commonly found at autopsy and may be seen in 20 to 27% of carefully examined pituitaries (12,13). Prolactin-producing and nonfunctioning (null cell) adenomas are the most commonly found incidental adenomas at autopsy.

Clinical Features

There is considerable age and sex variation in the incidence of various types of pituitary adenomas. Prolactinomas are most common in young women between 20 and 25 years of age. Growth hormone adenomas are seen in younger and older patients and have equal frequencies of densely and sparsely granulated tumors, although patients with sparsely granulated tumors are usually younger and have more recurrences of their tumors. Functioning ACTH cell adenomas have an unusually high female preponderance, whereas silent ACTH cell adenomas occur somewhat more frequently in men (30). Gonadotropic and nonfunctioning adenomas occur most commonly in patients older than 40 years. Null cell adenomas are the second most common type of tumor after prolactinomas. Recent evidence suggest that null cell adenomas and gonadotropic adenomas may be closely related (26).

Pituitary adenomas are relatively uncommon in childhood (59). Most of these childhood neoplasms are functional and are associated with gigantism and/or hyperprolactinemia.

Patients with pituitary adenomas usually present with signs and symptoms due to the hormone that is produced in excess by functional adenomas or with visual disturbances, impotence, headaches, and related signs and symptoms from a space-occupying lesion (Table 2.3). Slightly increased levels of serum prolactin may be associated with silent adenomas, and this is usually attributed to compression of inhibitory dopamine nerve axons by large tumors that lead to increase prolactin secretion. Immunohistochemical analysis in most of these cases fails to show PRL-producing cells in the neoplasm (17).

Table 2.3. Frequency and clinical signs and symptoms of pituitary adenomas

Tumor type	Percent of cases[a]	Major clinical signs/symptoms
Densely granulated GH adenoma	6.7	Acromegaly or gigantism
Sparsely granulated GH adenoma	7.3	Acromegaly or gigantism
Densely granulated PRL adenoma	0.6	Amenorrhea/ galactorrhea[c]
Sparsely granulated PRL adenoma	26.6	Amenorrhea/ galactorrhea,[c] impotence in men
Mixed GH-PRL adenoma	4.8	Acromegaly/ gigantism, hyperprolacnemia
Acidophil stem cell adenoma	2.2	Hyperprolactenemia ± acromegaly
Mammosomatotroph cell adenoma	1.4	Acromegaly/hyperprolactenemia
Functioning corticotroph cell adenoma	8.0	Cushing's disease, Nelson's syndrome
Silent corticotroph cell adenoma	6.0	Variable[b]
Thyrotroph cell adenoma	1.0	Hyper- or hypothyroidism
Gonadotroph cell adenoma	6.4	Variable[b]
Null cell adenoma	16.3	Variable[b]
Oncocytoma	8.9	Variable[b]
Plurihormonal adenoma	3.7	Variable[b]

[a]From 1043 biopsies in unselected surgical material. The percentage of cases is based on data from Horvath E, Kovacs K: Pituitary gland. Pathol Res Pract *183*: 129–142, 1988.
[b]Patients may be asymptomatic or have symptoms from large adenomas.
[c]Hyperprolactinemia commonly associated with amenorrhea and/or galactorrhea in premenopausal women.

Specific Adenoma Types

Pituitary adenomas have been classified historically as acidophils, basophils, and chromophobes. With the development of sophisticated diagnostic techniques such as immunochemistry and in situ hybridization, these neoplasms can be more precisely classified according to the hormone(s) they are producing. Pituitary adenomas can be divided into microadenomas (less than 10 mm in diameter) and macroadenomas (equal to or greater than 10 mm in diameter). A more complete classification by level of invasiveness as described by Hardy (21) is commonly used. Grade I tumors are confined to the sella turcica and are less than 10 mm in size. Grade II adenomas are larger than 10 mm but are still intrasellar. There is sellar enlargement but no destruction of bony structure. Grade III tumors cause localized erosion of the sella, while grade IV tumors are invasive adenomas with extensive destruction of bony structures.

Growth Hormone Adenomas

These tumors make up 8 to 14% of surgically resected adenomas. They are commonly acidophilic on H & E staining but may also be chromophobic. Nuclear pleomorphism may be seen in the benign neoplasms. These tumors are rarely cystic. Immu-

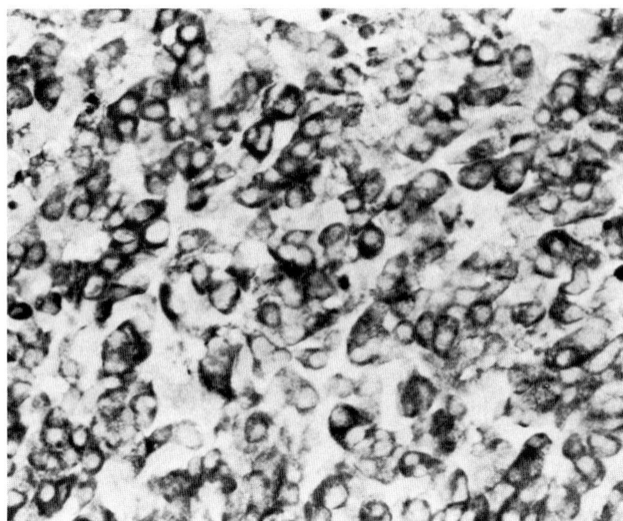

Figure 2.12. Immunohistochemical staining of a GH adenoma showing most of the tumor cells positive for GH.

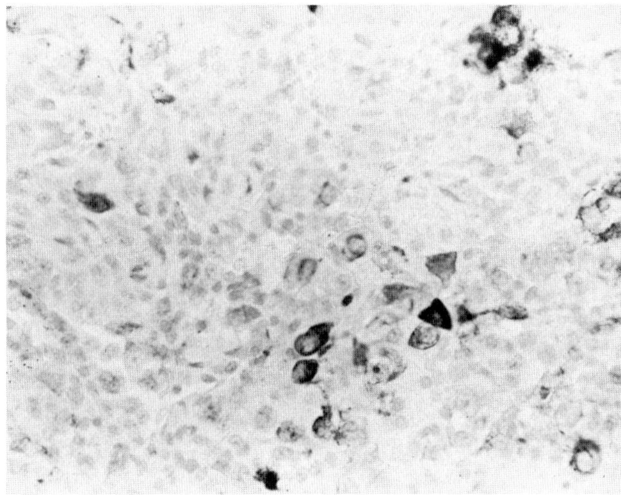

Figure 2.13. Some of the tumor cells in the GH adenoma from Fig. 2.12 stain positively for the alpha subunit of the glycoprotein hormones.

nostaining shows diffuse positive immunoreactivity for GH (Fig. 2.12) and many adenomas may also show immunoreactivity for PRL and for alpha subunit of the glycoprotein hormones (Fig. 2.13). The densely granulated GH adenomas have secretory granules 250 to 600 nm in diameter (Fig. 2.14), while the sparsely granulated tumors have smaller secreting granules 100 to 250 nm in diameter. Fibrous bodies are associated with sparsely granulated GH adenomas. These are spherically arranged aggregates of type II microfilaments that are usually present in the Golgi region.

Patients who have been treated with somatostatin or long-acting somatostatin analogs often have morphological changes in their tumors at surgery. There may be a slight to significant reduction in tumor size. Perivascular fibrosis and variable increases in the number of secretory granules have also been reported (7,35) (Fig. 2.15).

Many patients with acromegaly may have a mixed picture of GH-producing tumors. These include

1. A mixed growth hormone cell and prolactin cell adenoma in which separate cells produce each hormone (Fig. 2.16).

2. An acidophilic stem cell adenoma that is

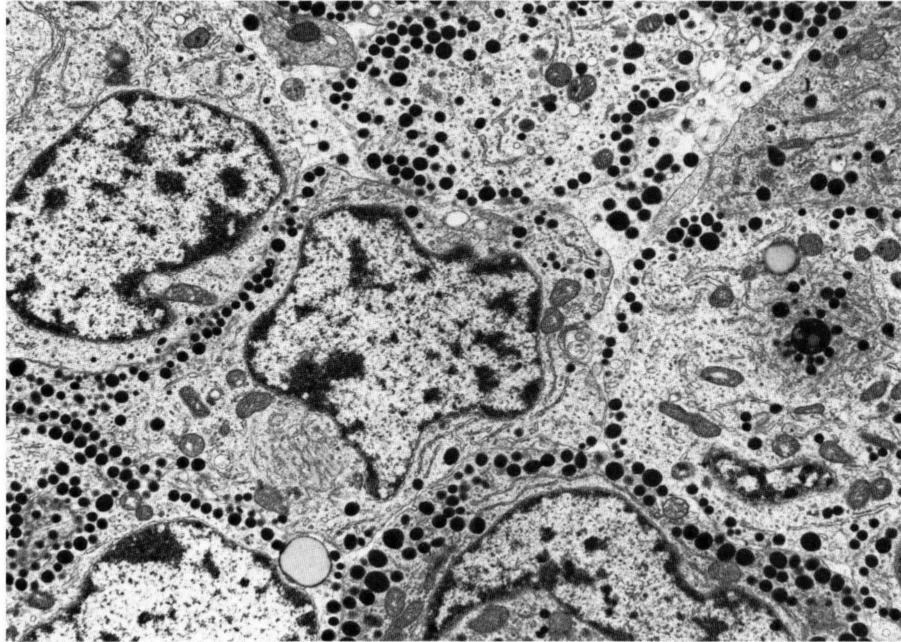

Figure 2.14. Densely granulated GH adenoma. The tumor cells have large secretory granules 250 to 600 nm in diameter (×6450).

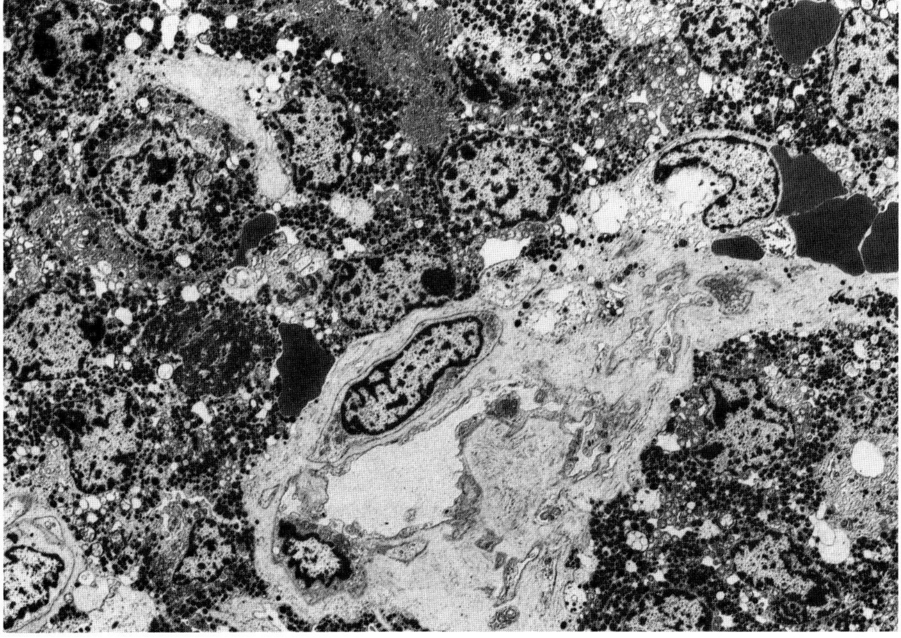

Figure 2.15. GH adenoma from a patient treated with long-acting somatostatin analog for 1 month before surgery. There is marked perivascular fibrosis (×2100).

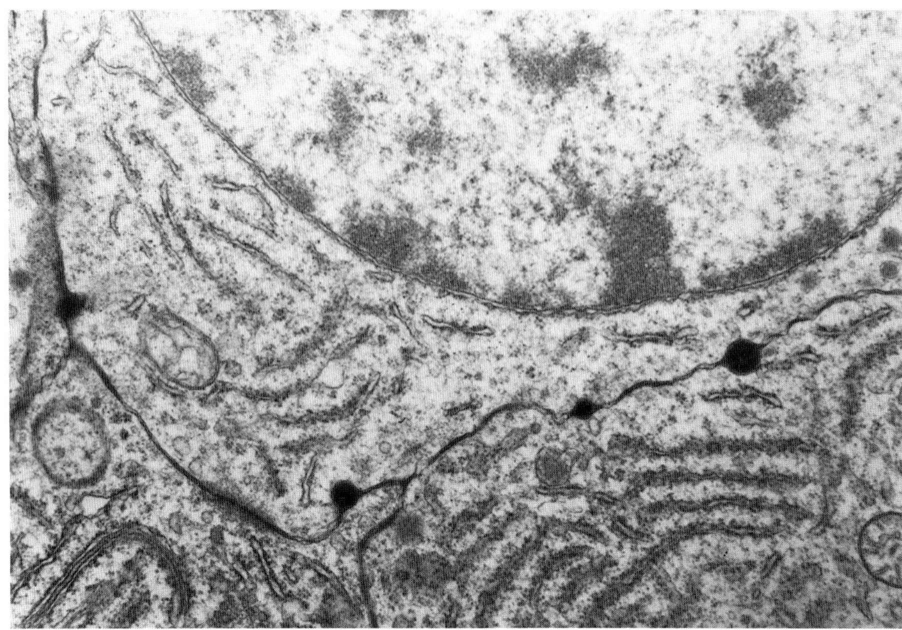

Figure 2.19. Ultrastructure of a sparsely granulated PRL adenoma showing reverse exocytosis (×20,625).

Corticotroph Cell Adenoma

Corticotroph cell adenomas may be functional or clinically silent (30). The functional tumors are associated with Cushing's or Nelson's syndrome (ACTH-secreting tumors in patients who have previously undergone bilateral adrenalectomies for preexisting Cushing's disease) and are much more common in women. Tumors associated with Cushing's disease are frequently microadenomas, while those with Nelson's syndrome tend to be larger and more aggressive. Some ACTH tumors may be only 1 to 2 mm in size. This small tumor may easily be lost during surgery or tissue processing and may explain why a small percentage of patients with Cushing's disease are cured by surgery even though a tumor is not seen by the pathologist. Sometimes sections through the entire embedded paraffin block may be needed to demonstrate the adenoma. On H and E sections the tumors range from chromophobic to basophilic. They are usually PAS positive. In general, chromophobic tumors are sparsely granulated. Immunochemistry shows ACTH immunoreactivity as well as positive staining for other proopiomelanocortin (POMC) cleavage products including endorphin and melanocyte-stimulating hormone (MSH). Ultrastructural examination of densely granulated tumors shows secretory granules 250 to 700 nm in diameter (Fig. 2.20). Sparsely granulated tumors have secretory granules between 100 and 200 nm in diameter. Variable amounts of type I microfilaments with a width of 7 to 10 nm are seen in adenomas associated with Cushing's disease. Crooke's hyaline change is more common in the nontumorous pituitary (Fig. 2.21) but may also be found in the adenoma cell. Microfilaments are uncommon in adenomas associated with Nelson's syndrome.

The silent corticotroph adenoma contains immunoreactive ACTH and related peptides from POMC proteolytic cleavage, but patients with these tumors do not have Cushing's disease (24). In situ hybridization studies have shown mRNA for ACTH, indicating active transcription of the POMC gene (R.V. Lloyd, unpublished observations). Some authors have divided these neoplasms into various subtypes based mainly on ultrastructural features (24). Subtype 1 is similar to the densely granulated ACTH adenoma on ultrastructural examination. They have type I microfilaments and increased

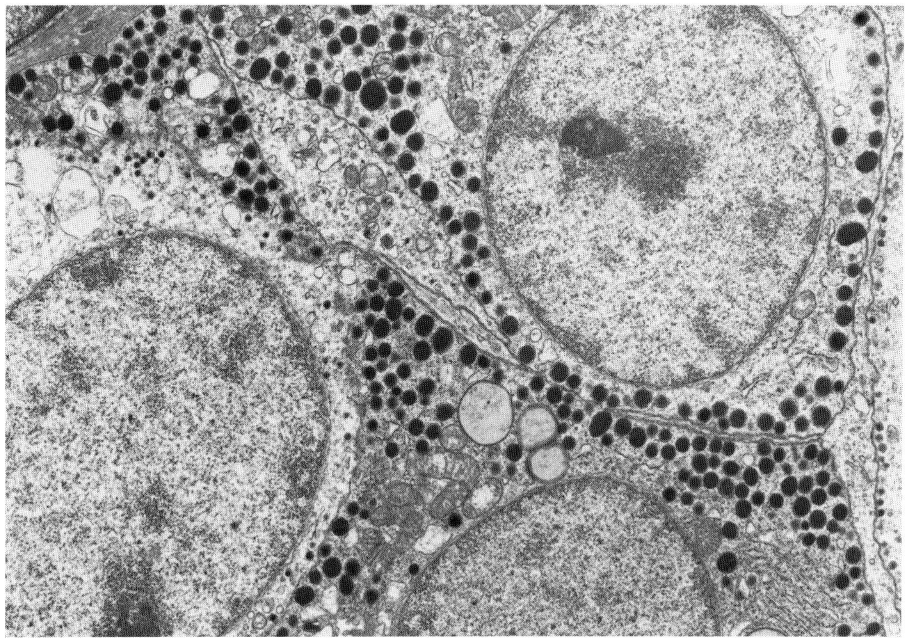

Figure 2.20. Electron micrograph of a densely granulated corticotroph adenoma with many secretory granules 250 to 700 nm in diameter (×6,600).

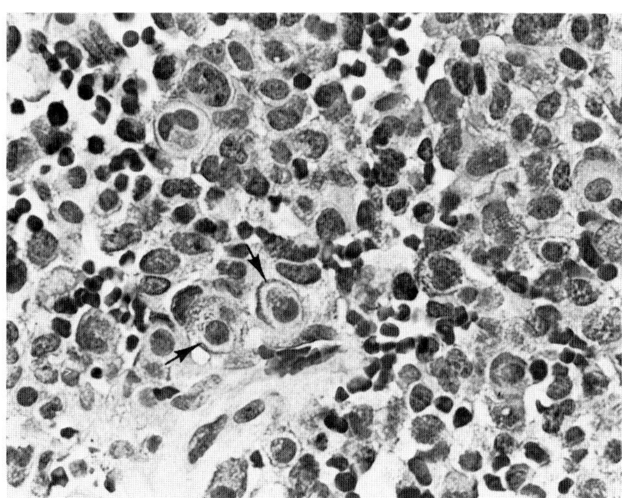

Figure 2.21. Prominent Crooke's hyaline change in the periadenomatous cells of the pituitary from a patient with Cushing's disease (*arrows*). The secretory granules are pushed toward the periphery of the cells and abundant microfilaments are present in the cytoplasms.

lysosomal activity. Subtype 2 has granules measuring 150 to 450 nm but no cytoplasmic microfilaments, while subtype 3 has small secretory granules up to 200 nm and many lysosomes. The subtype 3 tumors have been described as a new class of pituitary adenomas (25).

Gonadotropic Cell Adenomas

These tumors are commonly associated with elevated serum levels of FSH or LH or both and the alpha subunit of glycoprotein hormones. They are commonly chromophobic and PAS positive. Immu-

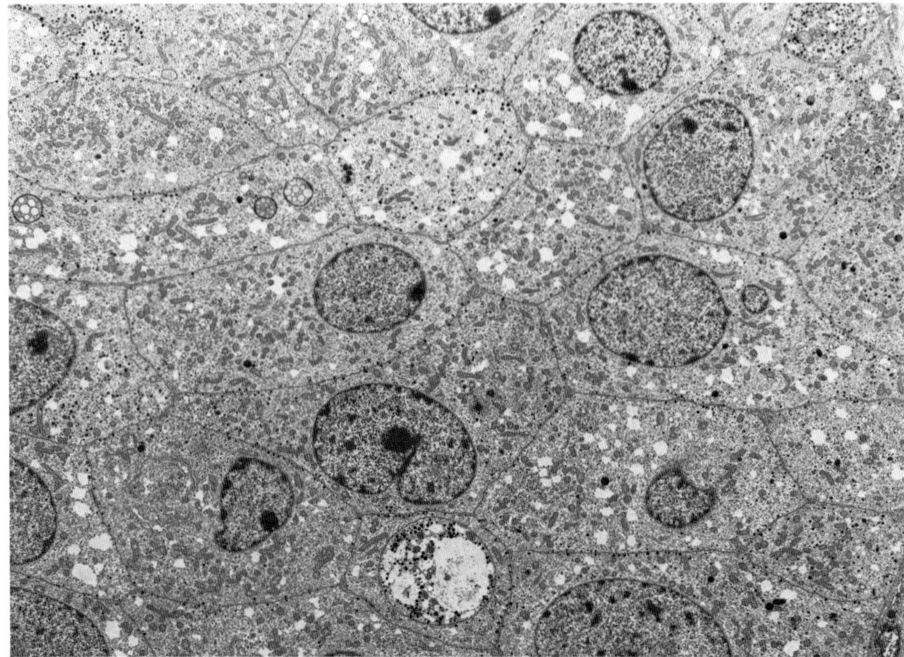

Figure 2.22. Gonadotropic adenoma consisting of elongated tumor cells. This tumor shows the male pattern of differentiation without prominent Golgi areas. The secretory granules range from 100 to 150 nm in diameter (×2184).

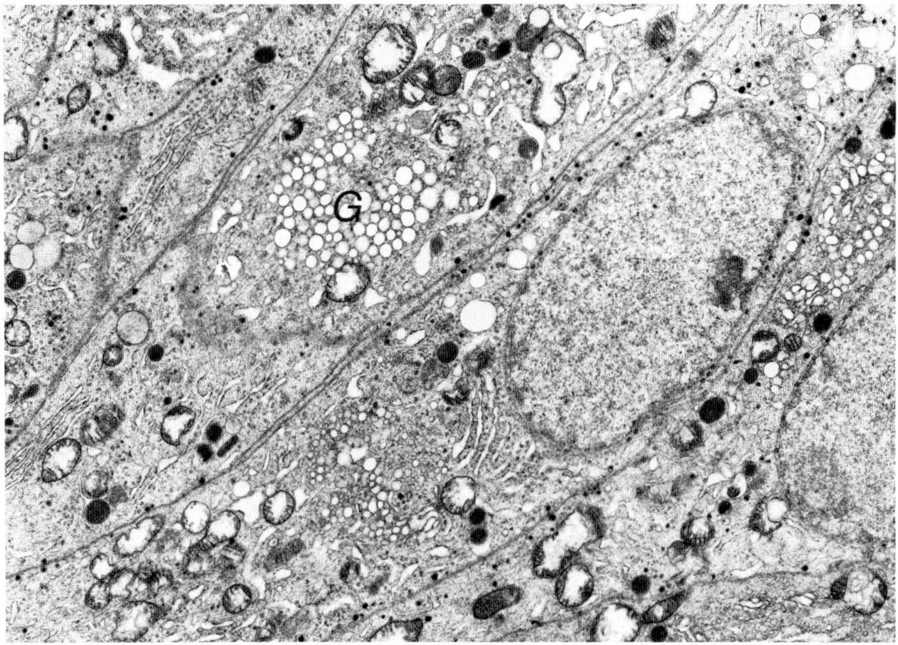

Figure 2.23. Gonadotropic adenoma from a female patient showing the female pattern of differentiation. The Golgi complexes (G) have a distinct honeycomb appearance. The secretory granules ranged from 100 to 200 nm in diameter (×6475). (Courtesy of Dr. K. Kovacs.)

nochemistry shows FSH and/or LH in the cytoplasm of variable numbers of tumor cells. Ultrastructural study shows a striking sexual dimorphism (30). The well-differentiated or "female-type" tumors consist of stacks of RER, rod-shaped mitochondria, and Golgi complexes with a honeycomb appearance. The secretory granules are less than 150 nm in diameter. Membrane-bound bodies, 400 to 450 nm, are usually present in the area of the Golgi complex. In the less differentiated or "male" pattern the RER is poorly developed and the Golgi complex is not prominent (Fig. 2.22), while in the female pattern of differentiation the Golgi complex has a distinct honeycomb pattern (Fig. 2.23). The secretory granules in both tumor types are small, less than 150 nm in diameter (75).

Thyrotroph Cell Adenoma

These are rare tumors that may be chromophobic or basophilic and are PAS positive. Immunostaining may be positive for TSH but is sometimes negative (Fig. 2.24). Ultrastructural studies reveal cells with small secretory granules, 100 to 250 nm, and

A

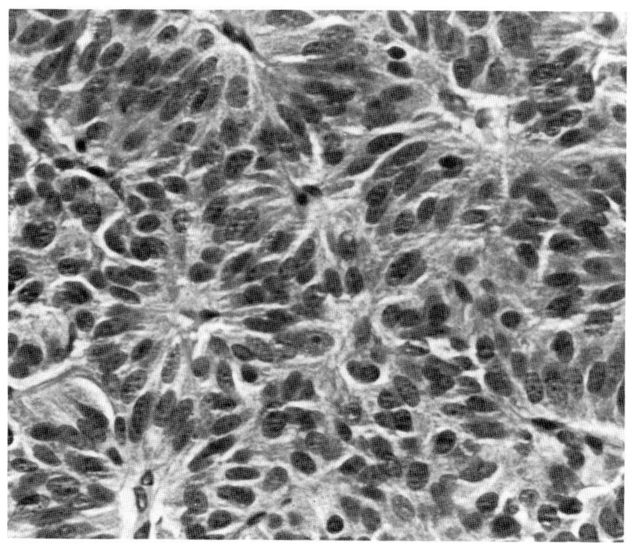

B

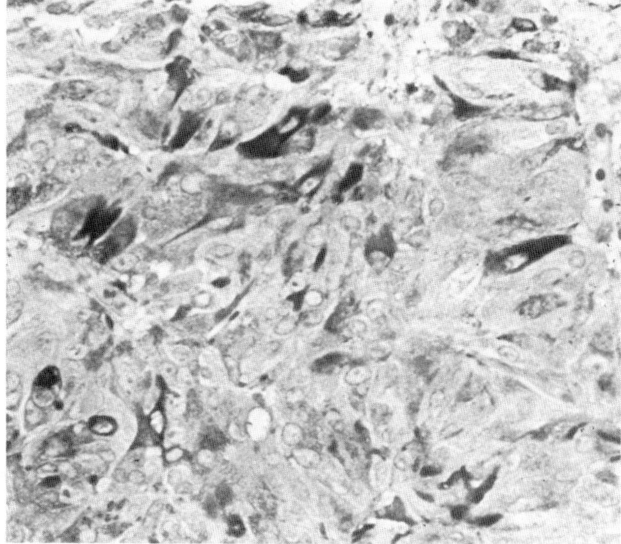

Figure 2.24. (A) Thyrotroph adenoma consisting of oval tumor cells. (B) Immunostaining for TSH showing strong positive immunoreactivity in some tumor cells. (C) Ultrastructure of a TSH cell adenoma showing small cells with prominent nucleoli and small secretory granules 100 to 250 nm in diameter. (Courtesy of Dr. K. Kovacs.)

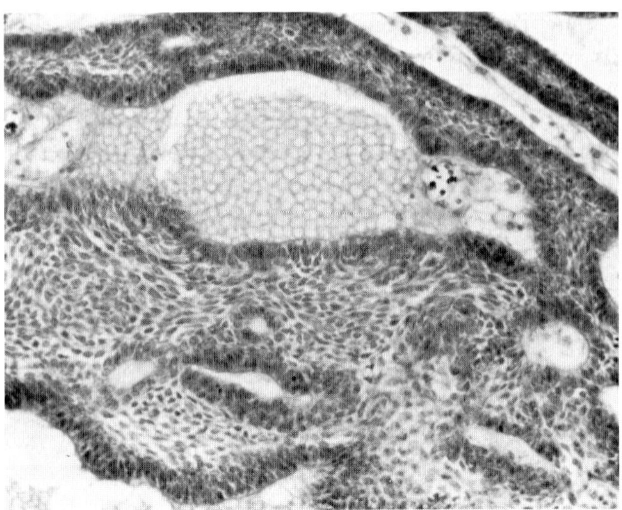

Figure 2.29. Craniopharyngioma with palisading basal cells, stellate cells of the epithelial component, and prominent cystic spaces.

tary hormones or neuroendocrine differentiation. Staining for keratin and other epithelial markers are usually positive.

Ultrastructural features include epithelial and connective tissue cells but no secretory granules or other evidence of neuroendocrine differentiation. Tonofilaments and desmosomal junctions are seen in the epithelial cells. These tumors are treated surgically, but complete removal is often difficult and recurrences are common. Postsurgical radiation therapy may be helpful.

The differential diagnosis includes other cystic lesions such as epidermoid cysts, Rathke's cyst, and cystic pituitary adenomas such as cystic prolactinomas, which can often be calcified.

Hypothalamic Gangliocytomas

These slow-growing lesions are also referred to as hamartomas, gliomas, and choristomas. They are histologically benign lesions that may represent either true neoplasms or congenital defects. The endocrine defects produced by gangliocytomas may be related to compression of the hypothalamus, pituitary, and/or hypophyseal stalk. Recent immunohistochemical studies have shown that gangliocytomas can secrete hypothalamic releasing hormones, resulting in endocrine symptoms. These lesions have been shown to secrete growth hormone releasing hormone, corticotropin releasing hormone, and gonadotropin releasing hormone.

Hamartomas occur in both males and females and may be found from infancy to adulthood but are more common in adults (3,4). Histologically, these lesions consist of randomly oriented large ganglion cells resembling those of the hypothalamus. Occasional binucleated neurons may be present (Fig. 2.30). Ultrastructural studies show neurons similar to those in the rest of the hypothalamus with neuronal processes, neurofibrillary material, and small dense-core secretory granules 100 to 300 nm in diameter. Immunochemical studies often reveal hypothalamic releasing tumors that may correspond to the patient's clinical signs and symptoms. Asa and her colleagues found growth hormone releasing hormone in the neurons of all six gangliocytomas examines in patients with hypothalamic acromegaly (4).

Most patients obtain symptomatic relief and deterioration of their clinical symptoms after surgical resection of the gangliocytomas (4).

Other Hypothalamic Diseases

Kallmann's Syndrome

Isolated gonadotropin deficiency can occur as a familial or a sporadic disease (28,36). It may be associated with anosmia, microphallus, and/or cryptorchidism. Failure to undergo puberty often leads to the discovery of the disorder. Partial defects in FSH and/or LH secretion may occur. The disease may be inherited as an X-linked or autosomal transmission with primary manifesta-

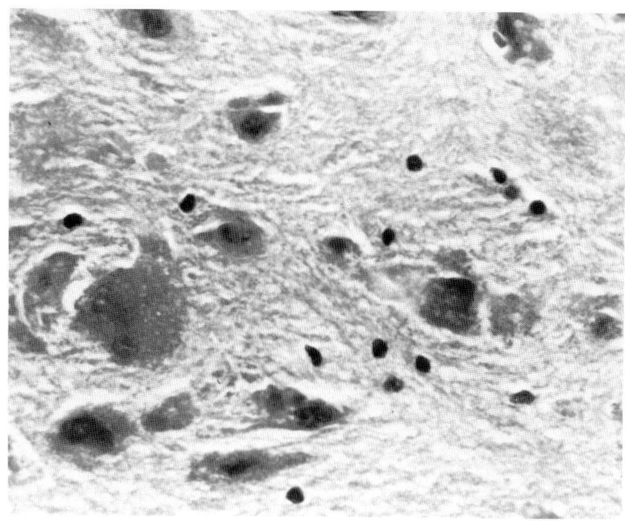

Figure 2.30. Hypothalamic hamartoma with prominent ganglion cells that are randomly oriented. These hamartomas frequently contain hypothalamic hormones and other neuropeptides.

tion in male offspring. Prolonged treatment with clomiphene may correct the defect in plasma LH. Administration of luteinizing hormone releasing hormone (LHRH) may also correct the hypothalamic defect.

Syndromes Associated with Secondary Obesity

Acquired obesity that develops after infancy may be associated with specific hypothalamic diseases. These include

1. The Laurence-Moon-Biedl syndrome, which is associated with obesity, retinitis pigmentosa, mental retardation, polydactyly, hypogonadism, and low gonadotropin levels (61).
2. Prader-Willi syndrome, associated with obesity, short stature, mental retardation, cryptorchidism, and small hands and feet. This disorder is associated with a deletion in chromosome 15 (37).

Anorexia Nervosa and Bulimia

These are common syndromes. Anorexia nervosa is associated with young women commonly between the ages of 12 and the early 30s. Patients are obsessed with the fear of being obese and hunger sensations are ignored or denied. It usually appears 4 to 5 years after menarche (67). It is characterized by cachexia, skin dryness, hypothermia, bradycardia, and hypotension. Endocrine abnormalities include amenorrhea and low FSH and LH values. However, with weight gain, reversal of the abnormalities often occurs. The rare cases of men with anorexia also have similar abnormalities and low serum testosterone levels. In spite of the hypothalamic dysfunctions, the primary disorder is probably psychiatric and treatment involves extensive psychiatric therapy.

Bulimia (voracious appetite or "ox hunger") is a syndrome characterized by enormous food intake over short periods of time (62). Many patients have a previous or present history of anorexia nervosa. The gorging is followed by induced vomiting and is commonly associated with the use of large amounts of laxatives. The syndrome is characterized by an irresistible urge to overeat accompanied by a fear of becoming obese. The physical signs often seen in anorexics are often absent in bulimia. Amenorrhea is present in a smaller percentage of patients and many patients are depressed. Endocrine abnormalities are less common in bulimia than in anorexia nervosa.

Miscellaneous Hypothalamic Disorders

Other hypothalamic disorders associated with endocrine manifestations that are not clearly defined include psychosocial dwarfism in children and maternal deprivation syndrome in infants with growth failure and defective growth hormone release. Interestingly, the latter disorder can be reversed and GH responses returned to normal by placing children in a more supportive environment (46).

Rare hypothalamic disorders such as periodic hypothalamic discharge of CRH may occasionally lead to Cushing's disease (77).

References

1. Asa SL, Bilbao JM, Kovacs, K, Josse RG, Kreines K (1981) Lymphocytic hypophysitis of pregnancy resulting in hypopituitarism: A distinct clinicopathologic entity. Ann Intern Med 95:166–171
2. Asa SL, Kovacs K, Bilbao JM (1983) The pars tuberalis of human pituitary. A histologic immunohistochemical, ultrastructural and immunoelectron microscopic analysis. Virchows Arch [Pathol Anat] 399:49–59
3. Asa SL, Kovacs K, Tindall GT, Barrow DL, Horvath E, Vicsei P (1984) Cushing's disease associated with intrasellar gangliocytoma producing corticotrophin releasing factor. Ann Intern Med 101:789–793
4. Asa SL, Scheithauer BW, Bilbao JM, Horvath E, Ryan N, Kovacs K (1984) A case for hypothalamic acromegaly: A clinicopathological study of six patients with hypothalamic gangliocytomas producing growth hormone-releasing factor. J Clin Endocrinol Metab 58:796–803
5. Baes, M, Allaerts W, Denef C (1987) Evidence for functional communication between folliculo-stellate cells and hormone-secreting cells in perfused anterior pituitary cell aggregates. Endocrinology 120:685–691
6. Baker BL, Jaffe RB (1975) The genesis of cell types in the adenohypophysis of the human fetus as observed with immunocytochemistry. Am J Anat 143:137–161
7. Barkan AL, Lloyd RV, Chandler WF, Hatfield MK, Gebarski SS, Kelch RP, Beitins IZ (1988) Preoperative treatment of acromegaly with somatostatin analog SMS 201-995: Shrinkage of invasive pituitary macroadenoma and improved surgical cure rate. J Clin Endocrinol Metab 67:1040–1048
8. Begeot M, Dubois MP, Dubois PM (1977) Growth hormone and ACTH in the pituitary of normal and anencephalic human fetuses: immunocytochemical evidence for hypothalamic influences during development. Neuroendocrinology 24:208–220
9. Bierich JR (1975) Sexual precocity. J Clin Endocrinol Metab 4:107–142
10. Boyd JD (1956) Observations on the human pharyngeal hypophysis. J Endocrinol 14:66–77
11. Branch CL Jr, Laws ER Jr (1987) Metastatic tumors of the sella turcica masquerading as primary pituitary tumors. J Clin Endocrinol Metab 65:469–474
12. Burrow GN, Wortzman G, Rewcastle NB, Holgate RC, Kovacs K (1981) Microadenomas of the pituitary and abnormal sellar tomograms in an unselected autopsy series. N Engl J Med 304:156–158
13. Costello RT (1936) Subclinical adenoma of the pituitary gland. Am J Pathol 12:205–216
14. Crooke AC (1935) A change in the basophil cells of the pituitary gland common to conditions which exhibit the syndrome attributed to basophil adenoma. J Pathol Bacteriol 41:339–349
15. Daniel PM, Prichard MML (1975) Studies of the hypothalamus and the pituitary gland. Acta Endocrinol (Kbh) 86 (Suppl 201):1–216
16. Ferrara N, Schweigner L, Neufeld G, Mitchell R, Gospodarowicz D (1987) Pituitary follicular cells produce basic fibroblast growth factor. Proc Natl Acad Sci USA 84:5773–5777
17. Frantz AG, Kleinberg DL, Noel GL (1972) Studies on prolactin in man. Recent Prog Horm Res 28: 527–573
18. Girod C, Trouillas J, Claustrat B (1986) The human thyrotropic adenoma: pathological diagnosis in five cases and critical review of the literature. Semin Diag Pathol 3:58–68
19. Goodyer CG, Guydan HJ, Giroud CJP (1972) Development of the hypothalamic-pituitary axis in the human fetus. In: Tolis G et al. (eds) Clinical Neuroendocrinology: A Pathophysiological Approach. New York, Raven Press, pp 199–214
20. Gooren LJG, Assies J, Asscheman H, DeSlegte R, Van Kessel H (1988) Estrogen-induced prolactinoma in a man. J Clin Endocrinol Metab 66:444–446
21. Hardy J (1973) Transphenoidal surgery of hypersecretory pituitary tumors. In: Kohler PO, Ross GT (eds) Diagnosis and Treatment of Pituitary Tumors. Int. Congress Series Nol. 303. Amsterdam: Excerpta Medica, pp 179–194
22. Hoi Sang U, Johnson C (1984) Metastatic prolactin-secreting pituitary adenoma. Hum Pathol 15:94–96
23. Horvath E, Kovacs K (1986) Pathology of prolactin cell adenomas of the human pituitary. Semin Diag Pathol 3:4–17
24. Horvath E, Kovacs K, Killinger DW, Smyth HS, Platts ME, Singer W (1980) Silent corticotropic adenomas of the human pituitary gland: A histologic, immunocytologic and ultrastructural study. Am J Pathol 98:617–638
25. Horvath E, Kovacs K, Smyth HS, Killinger DW, Scheithauer VW, Randall R, Laws ER Jr, Singer W (1988) A novel type of pituitary adenoma: morpho-

logical features and clinical correlations. J Clin Endocrinal Metab 66:1111–1118

26. Jameson JL, Klibanski A, Black MCL, Zerves NT, Lindell CM, Tsu DW, Ridgway EC, Habener JF (1987) Glycoprotein hormone genes are expressed in clinically non-functioning pituitary adenomas. J Clin Invest 80:1472–1478
27. Jordan RM, Kohler PO (1987) Recent advances in diagnosis and treatment of pituitary tumors. Adv Intern Med 32:299–324
28. Kallmann FJ, Shoenfeld WA, Barrera SE (1944) The genetic aspects of primary eunuchoidism. Am J Ment Defic 48:203–236
29. Kaufman B, Chamberlin WB Jr (1972) The ubiquitous "empty" sella turcica. Acta Radiol 13:413–425
30. Kovacs K, Horvath E (1986) Tumors of the Pituitary Gland. Atlas of Tumor Pathology, Second Series, Fascicle 21. Washington, DC, Armed Forces Institute of Pathology
31. Kovacs K, Horvath E, Ezrin C, Weiss MH (1982) Adenoma of the human pituitary producing growth hormone and thyrotropin: A histologic, immunocytologic and fine-structural study. Virchows Arch [Pathol Anat] 395:59–68
32. Kovacs K, Horvath E, Ryan N, Ezrin C (1980) Null cell adenoma of the human pituitary. Virchows Arch [Pathol Anat] 387:165–174
33. Kovacs K, Horvath E, Stratmann IE, Ezrin C (1974) Cytoplasmic microfilaments in the anterior lobe of the human pituitary gland. Acta Anat (Basel) 87: 414–426
34. Kovacs K, Lloyd RV, Horvath E, Asa SL, Stefaneanu L, Killinger DW, Smyth HS (1989) Silent somatotroph adenomas of the human pituitary. A morphologic study of three cases including immunocytochemistry, electron microscopy *in vitro* examination and *in situ* hybridization. Am J Pathol 134: 345–353
35. Landolt AM, Osterwalder V, Stuckman G (1987) Pre-operative treatment of acromegaly with SMS 201-995: surgical and pathological observations. In: Tolis G, Ludeke DK (eds) Growth Hormone. New York, Raven Press, pp 229–244
36. Laron Z, Kaushanski A, Josefsberg Z (1977) Penile size and growth in children and adolescents with isolated gonadotrophin deficiency (IGnD). Clin Endocrinol 6:265–270
37. Ledbetter DH, Riccardi VM, Airhart SD (1981) Deletions of chromosome 15 as a cause of the Prader-Willi syndrome. N Engl J Med 304:325–329
38. Linfoot JA (1979) Heavy ion therapy: Alpha particle therapy of pituitary tumors. In: Lindfoot JA (ed) Recent Advances in the Diagnosis and Treatment of Pituitary Tumors. New York, Raven Press, pp 245–267
39. Lloyd RV, Anagnostou D, Cano M, Barkan AL, Chandler WF (1988) Analysis of mammosomatotropic cells in normal and neoplastic human pituitary tissues by the reverse hemolytic plaque assay and immunocytochemistry. J Clin Endocrinol Metab 66:1103–1110
40. Lloyd RV, Cano M, Chandler WF, Barkan AL, Horvath E, Kovacs K (1989) Human growth hormone and prolactin secreting pituitary adenomas analyzed by *in situ* hybridization. Am J Pathol 134: 605–613
41. Lloyd RV, Chandler WF, Kovacs K, Ryan N (1986) Ectopic pituitary adenomas with normal anterior pituitary gland. Am J Surg Pathol 10:546–552
42. Lloyd RV, Chandler WF, McKeever PE, Schteingart DE (1986) The spectrum of ACTH-producing pituitary lesions. Am J Surg Pathol 10:618–626
43. McKeever PE, Koppelman MCS, Metcalf D, Quindlen E, Kornblith PL, Strott CA, Howard R, and Smith BH (1982) Refractory Cushing's disease caused by multinodular ACTH-cell hyperplasia. J Neuropathol Exper Neurol 41:490–499
44. Meichner RH, Riggio S, Manz JH, Earll JM (1987) Lymphocytic adenohypophysitis causing pituitary mass. Neurology 37:158–161
45. Melchionna RH, Moore RA (1938) The pharyngeal pituitary gland. Am J Pathol 14:763–771
46. Money J, Annecillo C, Kelley JF (1983) Growth of intelligence: failure and catchups associated respectively with abuse and rescue in the syndrome of abuse dwarfism. Psychoneuroendocrinology 8:309–319
47. Nakajima T, Yamaguchin H, Takahashi K (1980) S100 protein in folliculostellate cells of the rat pituitary anterior lobe. Brain Res 191:523–531
48. Neumann PE, Horoupian DS, Goldman JE, Hess MA (1984) Cytoplasmic filaments of Crooke's hyaline change belong to the cytokeratin class: an immunocytochemical and ultrastructural study. Am J Pathol 116:214–222
49. Nikitovitch-Winer BM, Atkin J, Maley BE (1987) Colocalization of prolactin and growth hormone within specific adenohypophyseal cells in male, female and lactating female rats. Endocrinology 121:625–631
50. Pearse AGE (1986) The diffuse neuroendocrine system: peptides, amines, placodes and the APUD theory. Prog Brain Res 68:25–31
51. Pelletier G, Robert F, Hardy J (1978) Identification of human anterior pituitary cells by immunoelectron microscopy. J Clin Endocrinol Metab 46:534–542

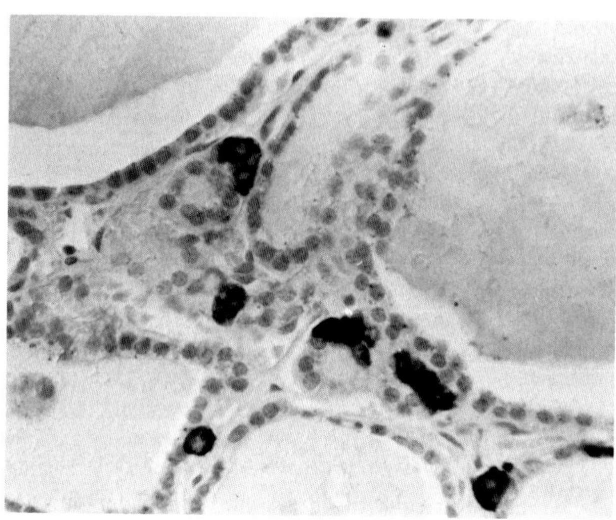

Figure 3.1. Calcitonin-producing C cells in the normal thyroid. The C cells are intrafollicular in location.

structural studies reveal dense-core secreting granules as in other neuroendocrine cells.

The thyroid is characterized by large stores of hormone and slow turnover of this hormone. Thyroglobulin is the storage form of the hormone. Triiodothyronine (T_3) and tetraiodothyronine (T_4) or thyroxine enter the blood directly after their cleavage from thyroglobulin by proteolytic enzymes within the follicular cells. Thyrotropin releasing hormone (TRH) from the hypothalamus regulates the release of thyroid-stimulating hormone (TSH) from the pituitary gland (Fig. 3.2). TSH stimulates the secretion of T_3 and T_4 and modulates changes within the follicular cells. T_3 and T_4 can regulate their secretion via a negative feedback effect on pituitary TSH and on hypothalamic TRH. Thyroid hormones have a plethora of effects on many metabolic processes, including stimulating calorigenesis, protein synthesis, and lipid metabolism. Thyroid hormones also affect many aspects of carbohydrate metabolism.

Abnormalities of the Thyroid

Developmental defects can range from complete absence of the thyroid, athyreosis, to failure of proper descent during embryonic development. Biosynthetic congenital metabolic defects in the thyroid, pituitary, or hypothalamus can all lead to hypothyroidism. Hypothyroidism in the newborn affects about 1/4000 to 1/5000 live births in the United States. Failure to recognize and treat this condition can result in cretinism (55) including irreversible mental retardation in infants. Pituitary or secondary hypothyroidism may result from postpartum pituitary necrosis or from a tumor of the pituitary or compression of the pituitary by adjacent lesions such as craniopharyngiomas.

The "lateral aberrant thyroid" is a historical observation which suggested that thyroid follicles located in a lateral position in the neck, including lymph nodes, represented ectopic thyroid follicles. This concept is incorrect (56) because most thyroid tissues outside the gland, including all thyroid tissues in lymph nodes, represent metastatic carcinoma until proved otherwise. Exceptions include 1) extension of thyroid tissue through the thyroid capsule into adjacent skeletal muscle, which is more prominent in young children and in patients with hyperplastic goiters as in Graves' disease; 2) ectopic tissue in the midline along the region of the thyroglossal duct; and 3) sequestered goiters arising from multinodular goiters that become completely detached from the thyroid. These sequestered nodules usually occur in older patients and can become calcified (66).

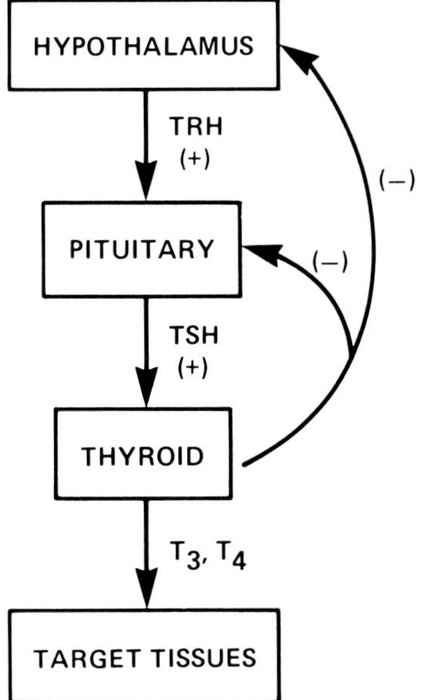

Figure 3.2. Feedback regulation of TSH and thyroid hormone secretion. The hypothalamus and other areas in the brain produce thyrotropin releasing hormone (TRH), which stimulates TSH production by anterior pituitary cells. TSH regulates thyroxine (T_4) and triiodothyronine (T_3) production by the thyroid gland. T_3 and T_4 have a plethora of effects on many metabolic processes. They also regulate secretion of TSH and TRH by a negative feedback mechanism.

Cyst

The thyroglossal duct cyst is the most common thyroid cyst (28). It represents vestiges of the primordial duct that have persisted as a benign chronic cyst. These cysts may communicate to the mouth via the foramen caecum. They are found in or near the midline of the neck usually just below the hyoid bone. The average size is 1 to 2 cm and they have a thin wall. Infected or recurrent cysts often have fibrotic walls. The content of the cyst is a clear mucinous fluid that can become purulent with infection. The microscopic appearance is that of a simple columnar lining, which is often ciliated but may be transitional, squamous, or mixed. Small foci of normal thyroid tissue are frequently seen in the cyst wall, but accumulations of lymphocytes are usually not present, which can help to distinguish thyroglossal duct cysts from branchial cleft cysts. Branchial cleft cysts are usually located in the lateral part of the neck.

Drugs Affecting Thyroid Morphology

Pigmented or black thyroid is a condition resulting from chronic minocyclin use (3). The gland is darkly pigmented on gross examination. Microscopic examination shows cytoplasmic dark brown pigment that is negative for iron. Ultrastructural studies show lipofuscin-type pigment in the thyroid follicles (60). This condition does not affect normal thyroid function.

Propylthiouracil (PTU), which blocks the conversion of iodide to iodine within the follicular cells and prevents synthesis of thyroid hormones, also stimulates release of pituitary TSH with proliferation of the thyroid epithelium. With PTU therapy in Graves' disease the epithelium remains columnar, the follicles are small, and colloid is scanty or absent. Treatment with iodine leads to involution of the follicles as they become filled with colloid. The epithelium becomes flattened and low cuboidal and the gland has decreased vascularity (73).

Thyroid Atrophy in Adults

Primary thyroid atrophy in adults is more common in women than in men, is most often seen between 40 and 60 years of age, and results in myxedema. The etiology is unknown. Circulating autoantibodies are seen in up to 80% of patients. It may represent the end stage of autoimmune thyroiditis in which a goiter was absent or went unnoticed (36). The gland is small and consists mostly of fibrous tissue with occasional thyroid follicles and lymphocytic infiltrates.

Thyroiditis

Inflammatory diseases affecting the thyroid can range from severe conditions such as Hashimoto's thyroiditis and subacute thyroiditis to lesions such as thyroiditis of pregnancy.

A

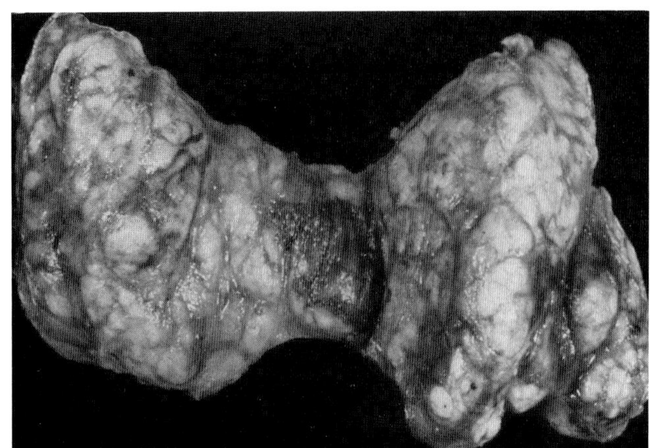

B

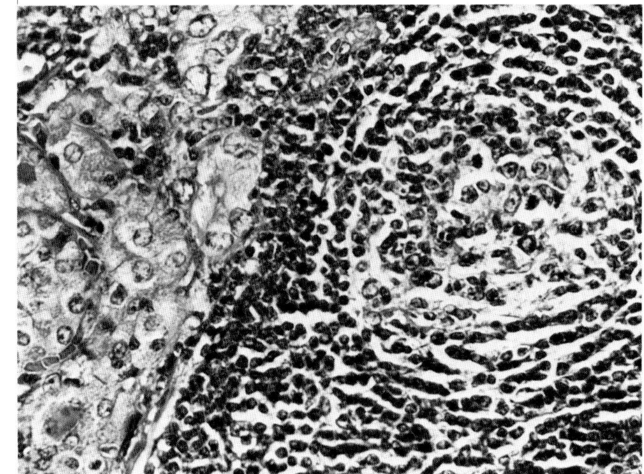

Figure 3.3. Gross appearance of thyroid in Hashimoto's thyroiditis (A). Multiple nodules are present in both lobes and in the isthmus. (B) Microscopic examination shows prominent lymphoid follicles with germinal centers and Hürthle or Askanazy cells, which are metaplastic follicular cells.

Hashimoto's Thyroiditis

This condition is also known as autoimmune thyroiditis, lymphocytic thyroiditis, or struma lymphomatosa. The patient usually has circulating antithyroglobulin and antimicrosomal antibodies. It affects women more commonly and often occurs between ages 30 and 50, but no age is exempt. Specific HLA-DR subtypes including HLA-DR3 and HLA-DR5 are more frequently associated with Hashimoto's disease. Possible mechanisms include humorally mediated autoimmunity, cell-mediated autoimmunity, and a combination of both leading to lymphocyte-mediated cytotoxicity that is targeted and triggered by antithyroid antibodies (74).

The thyroid gland is pale, nodular, and firm (Fig. 3.3A). Both lobes are affected, but nodules may be more prominent in one lobe. Microscopically, lymphocytic infiltrates admixed with plasma cells are seen. Prominent germinal centers and Hürthle or Askanazy cells are present (Fig. 3.3B). The degree of fibrosis is related to the chronicity of the disease. Squamous metaplasia may be present. Because chronic lymphocytic thyroiditis can be seen frequently in the thyroid gland, clinicopathologic correlation including elevated antibody titers should be used to make the diagnosis of Hashimoto's or autoimmune thyroiditis in cases in which not all of the histologic features are present. Lymphomas arising in the thyroid often develop in a background of Hashimoto's disease, although most patients with Hashimoto's disease do not develop thyroid lymphomas. The fibrous variant of Hashimoto's disease has many features of Hashimoto's thyroiditis but can have a confusing histological picture (38). The gland is enlarged, but extensive

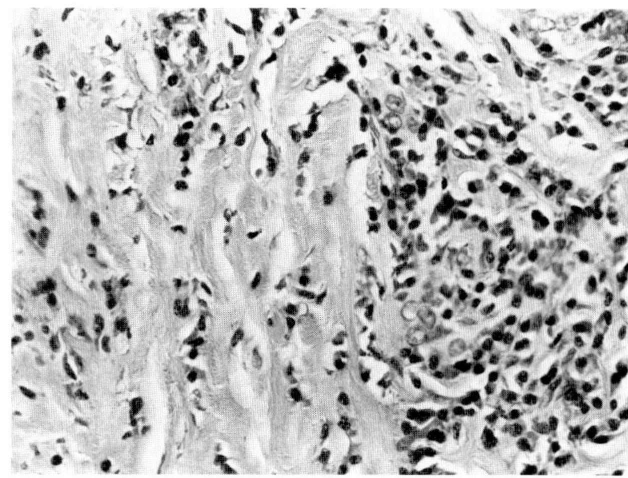

Figure 3.4. Fibrous variant of Hashimoto's thyroiditis. Extensive areas of fibrosis with collagenous bands and a mixed lymphocytic and plasmacytic inflammatory infiltrate are present.

fibrosis and less inflammation are seen microscopically (Fig. 3.4). Squamous metaplasia is abundant. Residual follicles entrapped in a fibrous background may appear infiltrative, giving the histopathological impression of an invasive follicular carcinoma. These patients usually have very high titers of antithyroglobulin antibodies (38).

Chronic Lymphocytic Thyroiditis

Some thyroid glands may have a lymphocytic infiltrate but do not have germinal centers or Askanazy cells as seen with Hashimoto's thyroiditis. The antithyroid antibody titers may be normal to slightly elevated. These findings are not specific and the diagnosis of chronic lymphocytic thyroiditis is given to these cases (43).

Subacute Thyroiditis

This condition, also known as granulomatous giant cell or de Quervain's thyroiditis, is usually caused by a viral infection of the thyroid gland and frequently follows an upper respiratory illness. The clinical picture of a painful thyroid associated with fever and elevated sedimentation rate is so classical that biopsies or cytologic aspirations are rarely done.

The histopathologic picture shows patchy areas of mononuclear and giant cells infiltrating the follicles with disruption of the epithelium and partial or complete loss of colloid (Fig. 3.5A and B). The follicular changes may be associated with some interfollicular fibrosis and interstitial inflammation. When the disease subsides, the histopathological appearance of the gland returns to normal.

The differential diagnosis of granuloma in the thyroid would include sarcoidosis (Fig. 3.6), in which the granulomas are mostly in the interstitium, fungal diseases, and tuberculosis involving the thyroid with diffuse or focal granulomatous inflammation. "Palpation thyroiditis" is often a focal lesion consisting of histologic cells forming granulomas in response to vigorous clinical palpation of the thyroid (10).

Riedel's Thyroiditis

This is a rare condition that is seen most commonly in middle-aged women (65,79). It may be associated with other fibroinflammatory conditions including retroperitoneal fibrosis and sclerosing mediastinitis. The gland may be of normal size or enlarged. There is a dense fibrosis replacing follicles and extending into the adjacent soft tissues of the neck. Microscopically, the thyroid parenchyma is replaced by dense collagen with chronic inflammatory cells. The patient may have dysphagia, dyspnea, and hoarseness but is usually euthyroid early and subsequently becomes hypothyroid as more of the thyroid is replaced. Riedel's may be confused with carcinoma because of the follicular cells entrapped in the scar. It may also look like the fibrous variant of Hashimoto's disease histologically, but the antibody titers are usually only slightly elevated. Schwaegerle and her col-

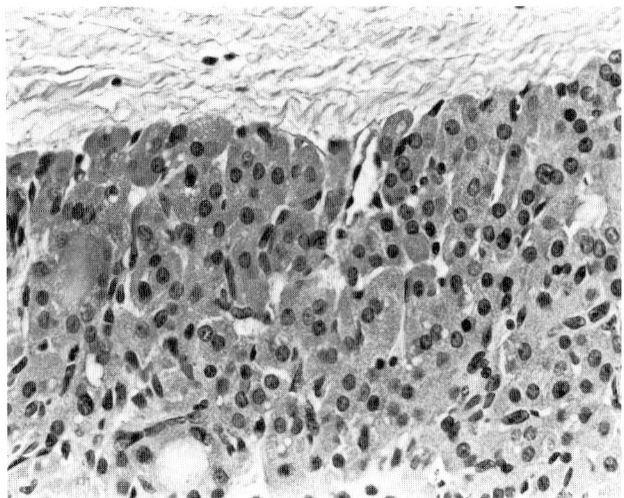

Figure 3.13. Hürthle cell adenoma. The neoplastic cells have large granular eosinophilic cytoplasm. There was no evidence of vascular or capsular invasion in this tumor.

various patterns of adenomas, the embryonal and fetal patterns are usually more cellular. Careful examination for vascular and capsular invasion to exclude the diagnosis of a carcinoma should be done with all adenomas.

Other Adenomas

The Hürthle cell adenoma (5) is a variant of follicular adenomas. It is composed predominantly of cells with granular bright eosinophilic cytoplasm that contain many mitochondria on ultrastructural examination (Figs. 3.13, 3.14).

Hyalinizing trabecular adenomas (Fig. 3.15) are benign neoplasms that are sometimes misdiagnosed as carcinomas (11). The adenomas are more common in women and occur between 27 and 72 years of age with a mean of 46 years. The tumors range from 0.3 to 4 cm and are tan-yellow and encapsulated. Microscopically, the cells are polygonal to oval and elongated, arranged in trabeculae with fine granular cytoplasm. Cytoplasmic invaginations into the nucleus and nuclear grooves may be present. Occasional psammoma bodies and perivascular hyaline fibrosis can be seen. Immunochemistry shows positive immunoreactivity for thyroglobulin and negative staining for calcitonin (11).

Atypical adenomas are usually composed of spindle to oval cells with irregular nuclei and some degree of pleomorphism and mitotic activity. Few follicles are present (32). Although these lesions may resemble carcinomas, invasion is not seen after extensive sampling and the tumors do not metastasize even after long clinical follow-up (32).

The differential diagnosis of atypical adenomas includes medullary thyroid carcinomas and solitary foci of metastatic carcinoma to the thyroid. Immunohistochemical stains for calcitonin, thyroglobulin, and other epithelial markers are helpful in establishing the diagnosis. Multiple sections through the capsule with a minimum of 10 sections of the entire nodule, if the entire lesion is too large to be examined in toto, should be submitted in the evaluation of a follicular nodule for evidence of malignancy (42).

Other uncommon benign tumors of the thyroid include teratomas (57), granular cell tumors, paragangliomas, and hemangiomas.

Malignant Thyroid Neoplasm

Papillary Neoplasms

Clinical Features

Papillary thyroid carcinomas occur two to three times more commonly in women than in men. A history of ionizing radiation to the neck region has been implicated with an increased incidence of papillary carcinoma of the thyroid (29,33). Papillary carcinoma is the most common subtype in adults and in children (71,76). Many childhood papillary carcinomas have been associated with previous head and neck irradiation, usually of 180 to 6000 roentgens with a average of about 600 roentgens

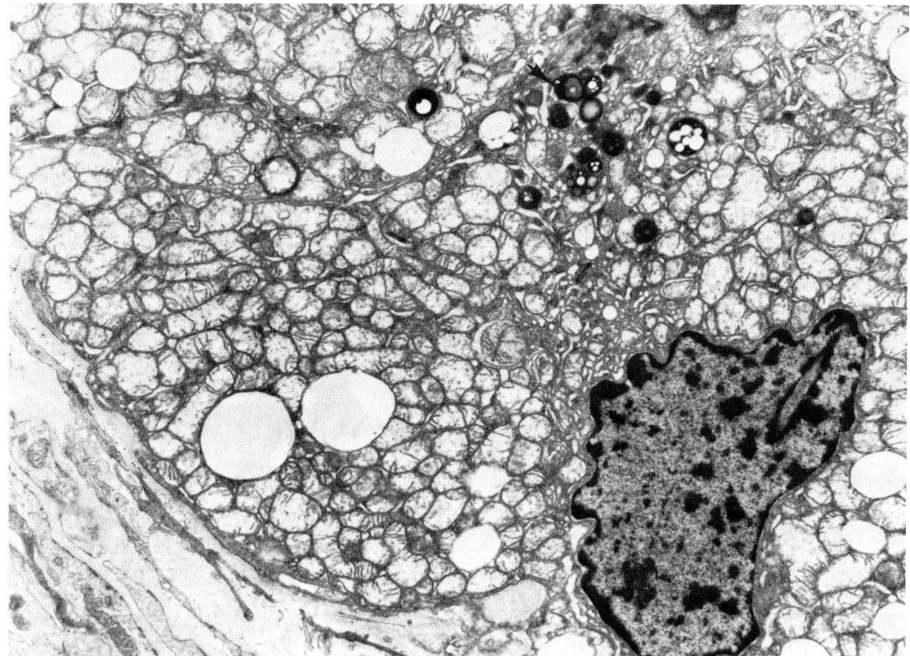

Figure 3.14. Ultrastructural examination of a Hürthle cell adenoma showing many mitochondria and a cluster of lysosomes (*arrow*) (×6156).

(41). The latency period for development of carcinoma ranges from 3.6 to 14 years with an average of 9 years. Hot thyroid nodules are rarely malignant; most thyroid neoplasms present as cold nodules. However, most "cold" nodules are benign lesions. Papillary carcinomas commonly present as asymptomatic thyroid nodules or as an enlargement of the regional lymph nodes. Thyroid neoplasms occurring in patients less than 40 to 50 years of age generally have a better prognosis than those in older patients (17).

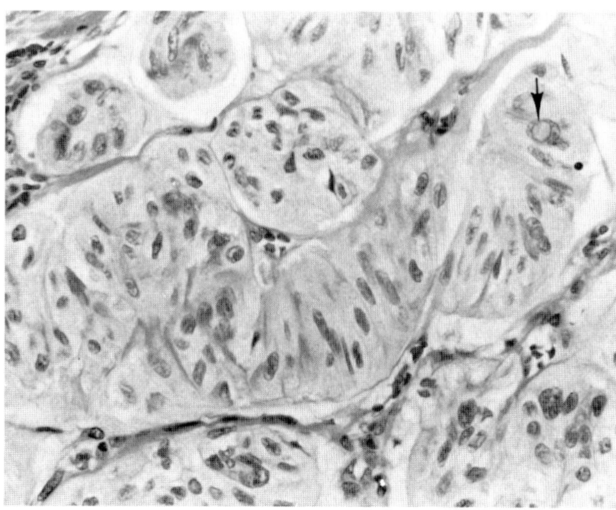

Figure 3.15. Hyalinizing trabecular adenoma. This benign neoplasm consists of polygonal to oval cells with abundant granular cytoplasm and cytoplasmic invagination into the nucleus (*arrow*).

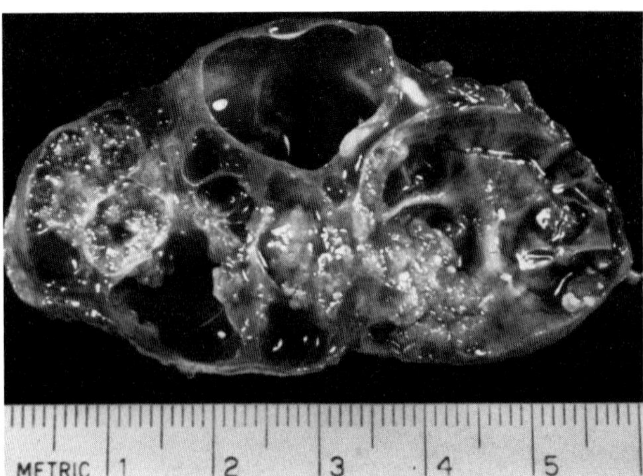

Figure 3.16. Gross appearance of a papillary thyroid carcinoma with cystic and solid areas. Focal fibrosis is also present.

Pathology

The gross appearance of papillary carcinomas varies with the size of the tumors. Large tumors usually have poorly defined borders, and some tumors may be partially encapsulated. Cystic change is common in large tumors. The cyst may be hemorrhagic or filled with brown watery fluid. Portions of the cyst may have a smooth lining while in other areas papillae project from the inner lining (Fig. 3.16). Central dense scars and fibrosis are commonly seen even in small, minimal, or occult carcinomas less than 1.0 cm in diameter. Calcification is frequently present and ossification is occasionally seen.

The microscopic appearance of papillary thyroid carcinomas includes true papillae with fibrovascular cores (Fig. 3.17A and B). The papillae may be present only focally or may be seen in most of the tumor, with the former exemplifying the follicular variant of papillary carcinoma (70) (Fig. 3.18). The nuclei are irregular, large, and overlapping or crowded and have a pale ground-glass appearance. Pseudoinclusions or cytoplasmic invaginations into the nucleus are often present. The ground-glass appearance and pseudoinclusion produce the clear round space or "Orphan Annie" nuclei. The ground-glass nuclei are difficult to appreciate on frozen sections. However, cytoplasmic invagination into the nucleus can be seen on frozen sections.

Psammoma bodies are present in about 40% of cases. True psammoma bodies have distinct concentric laminations, while pseudopsammoma bodies are calcific foci that are often stellate and irregular (70). Psammoma bodies are more common in patients less than 20 years of age with papillary carcinoma. Their presence is highly correlated with lymph node metastasis, but they have no other proven clinical or pathologic correlation. The stroma of papillary carcinomas is usually desmoplastic and mitoses are relatively uncommon.

Variants of Papillary Carcinomas

Follicular Variant

This lesion was first described by Lindsay (44) and subsequently by others (13). The tumor consists almost entirely of follicles with large cells with ground-glass overlapping nuclei with nuclear grooves (Fig. 3.18). They may or may not be encapsulated. Extensive sampling often reveals focal areas with papillae.

Encapsulated variants of papillary carcinoma are rare. They may coexist with different patterns including transitional and glandular areas (71). They usually have an excellent prognosis (22).

Diffuse Sclerosing Variant

These unusual tumors consist of diffuse involvement of one or both lobes by dense sclerosis with abundant psammoma bodies mixed with atypical papillary carcinoma elements (Fig. 3.18B and C). Foci of squamous metaplasia are present and should be distinguished from squamous carcinoma (70).

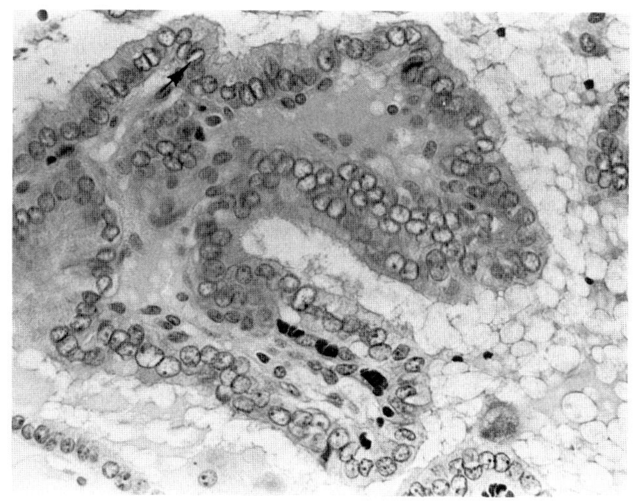

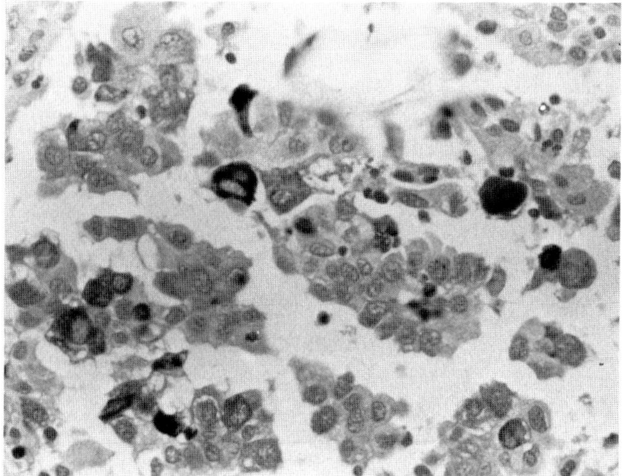

Figure 3.17. Papillary thyroid carcinoma. (A) The tumor cells have enlarged overlapping nuclei and longitudinal nuclear grooves (*arrow*), and there is a central fibrovascular core in the papillae (B). Immunostaining for thyroglobulin is positive in some tumor cells of a papillary carcinoma.

Columnar Cell Carcinoma

These rare carcinomas are highly aggressive papillary carcinomas composed of papillae lined by cells with marked nuclear stratification. Solid foci of spindle cells and microfollicular regions are also present (21). These tumors usually have a poor prognosis (21).

Tall Cell Variant

This rare papillary carcinoma was originally described by Hawk and Hazard (30,37). These tumors are usually large and bulky with abundant eosinophilic cytoplasm. The most diagnostic features are the tall columnar cells with basally located nuclei (unlike the cells in columnar cell carcinoma) and with a height that is at least twice the width (Fig. 3.19). A report by Johnson and coworkers (37) indicated that this histology was associated with a worse prognosis regardless of the age of the patient. In their study at least 30% of the tumor was required to have this histology before the diagnosis was made.

Minimal or Occult Papillary Carcinoma

These are small carcinomas that are not clinically apparent. Patients may present with lymph node or other metastasis as the first indication of thyroid carcinoma. Although the size was originally desig-

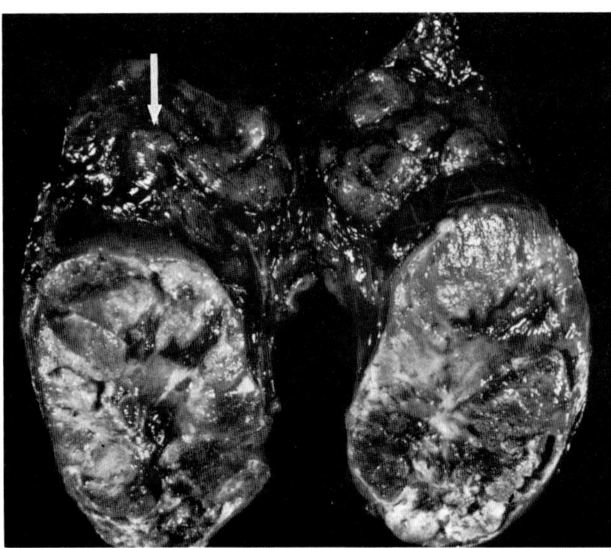

Figure 3.21. Gross appearance of a follicular carcinoma. This tumor arose in a thyroid with a multinodular goiter (*arrow*) that was present for over 40 years. Areas of necrosis and hemorrhage are present.

are usually not present in follicular neoplasms. Expression of HLA-DR is commonly seen in papillary but not in follicular carcinomas (46). Medullary thyroid carcinomas can look like papillary carcinomas because of the presence of fibrosis and calcification. Staining for amyloid and immunohistochemical staining for thyroglobulin and calcitonin help to separate these two lesions. When a papillary neoplasm is present at a distant site such as the lungs, thyroglobulin immunostaining is the most definitive marker to distinguish between a thyroid primary and an ovarian or other primary site since it is almost unique for thyroid epithelium (it may also be present in struma ovarii).

Papillary carcinomas commonly metastasize to lymph nodes and to lungs. These tumors occasionally invade intrathyroidal blood vessels.

Follicular Carcinomas

Follicular carcinomas constitute about 5 to 10% of thyroid carcinomas. They usually occur in an older age group than papillary carcinomas and are relatively uncommon before age 40. Like papillary neoplasms, follicular carcinomas occur two to three times more commonly in women than in men. A history of head and neck irradiation may be present in a few cases. The patient may have had a goiter for 10 years or longer before the development of the malignant neoplasm. Patients usually present with a thyroid nodule or mass that is quite firm. It may involve part of the lobe or the entire lobe. Pain is usually not present early in the course of the disease. Occasionally patients may present with pulmonary metastasis or a pathological fracture as the first manifestation of follicular carcinoma.

Pathology

Follicular carcinoma is usually fleshy and may have a fibrous center. The tumors may vary from 1 to 10 cm or greater in size. Retrogressive changes including hemorrhage, fibrosis, infarction, and calcification may be present (Fig. 3.21). Grossly, the tumors may show extensive invasion of the adjacent thyroid, blood vessels, and adjacent structures.

Microscopic examination shows two basic patterns of follicular carcinomas: encapsulated and widely invasive (16,23,24,26,77). Encapsulated carcinomas are tumors that are well encapsulated grossly but show microscopic evidence of capsular and/or vascular invasion (Fig. 3.22A–C). The widely invasive carcinomas include follicular carcinomas that are nonencapsulated and those that are encapsulated but have marked tissue or vascular invasion (23).

The criteria for malignancy in follicular neoplasms include vascular and/or capsular invasion

Figure 3.22. Follicular carcinoma showing variable patterns. The tumor cells have decreased colloid stores (A and B). A focus of angioinvasion is present within the capsule (C).

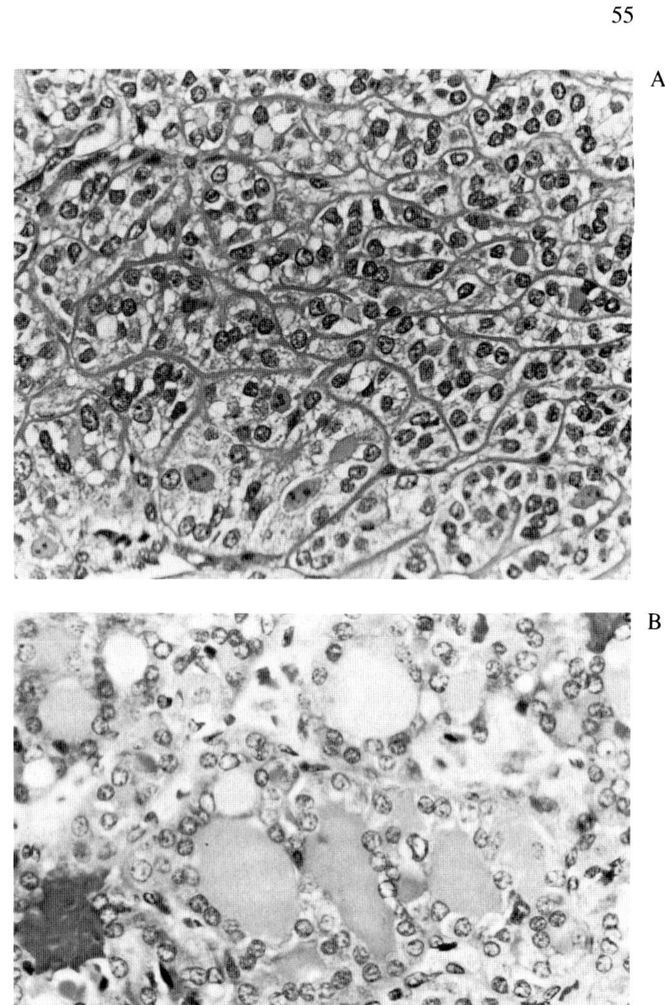

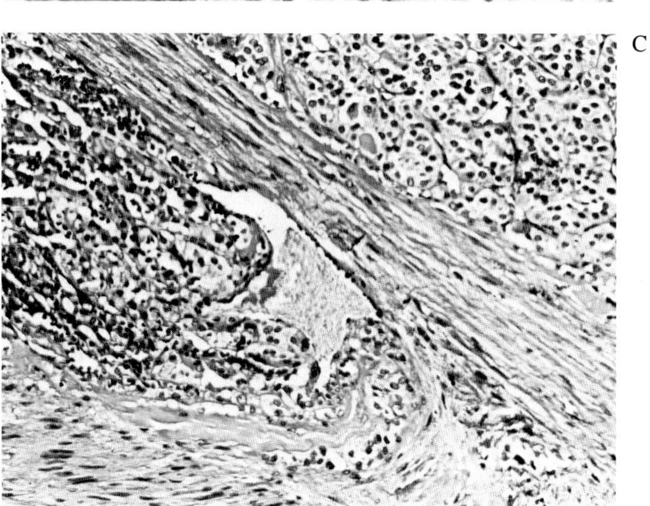

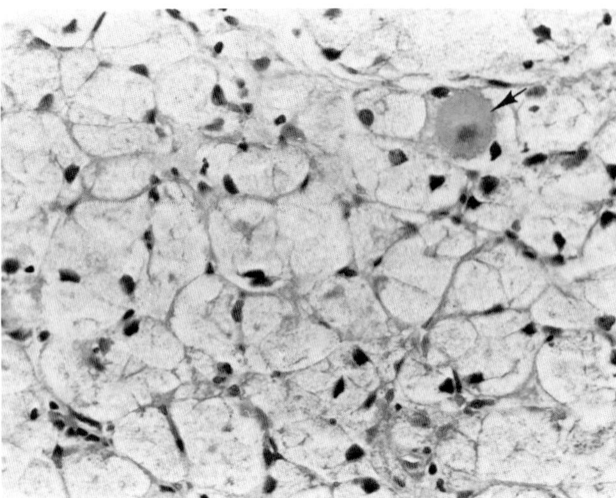

Figure 3.23. Clear cell follicular adenoma. The neoplastic cells have abundant clear cytoplasm. Evidence of thyroglobulin storage by the tumor cells is shown in one follicle (*arrow*).

(Fig. 3.22C). Vascular invasion should be present within or outside the capsule. In true vascular invasion, there are plaques of tumors forming polypoid projections into vessel lumina (23). Tumor thrombi covered by an endothelial layer or attached to the vessel wall are good evidence of vascular invasion. Capsular invasion should be diagnosed only if there is total penetration of the capsule by tumor tissue and adjacent thyroid tissue or other nonthyroid structures (23). The importance of obtaining a minimum of 10 sections through the capsule to evaluate for invasion in large follicular neoplasm has been emphasized by Lang and his colleagues (42). If the tumor shows marked cellularity, atypical histological features suggestive of invasion, additional sections should be taken if the initial screening did not show evidence of vascular invasion. Increased mitoses and nuclear atypia are not helpful features in the diagnosis of follicular carcinomas (23).

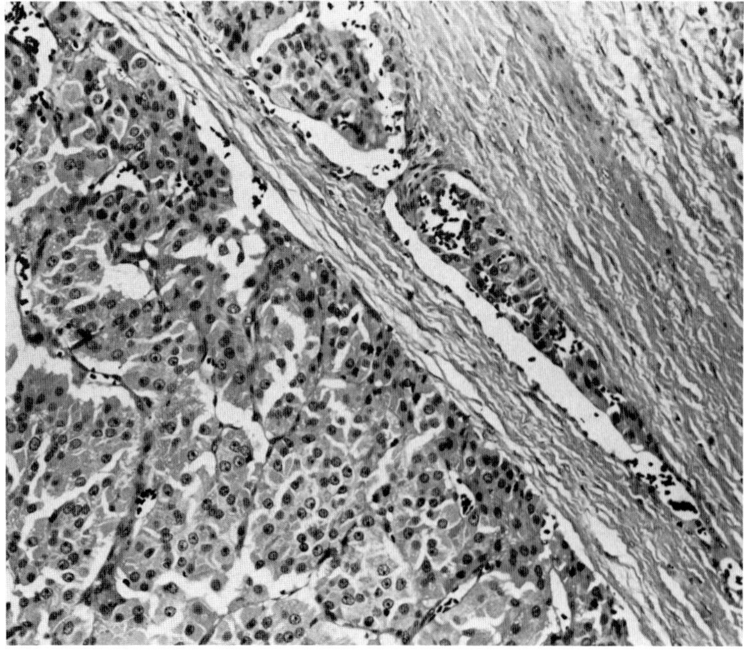

Figure 3.24. Hürthle cell carcinoma with foci of vascular invasion.

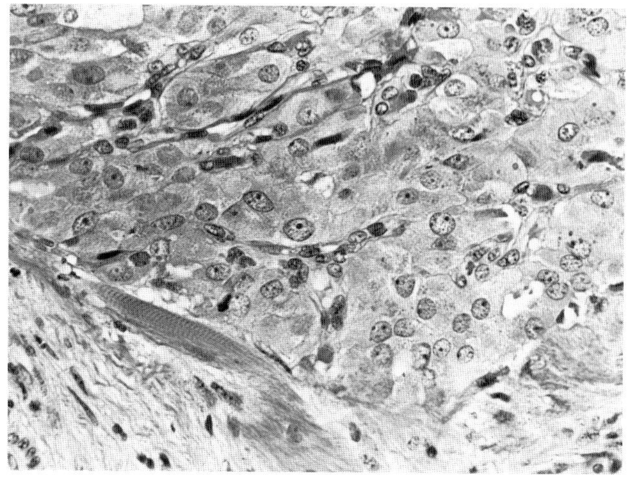

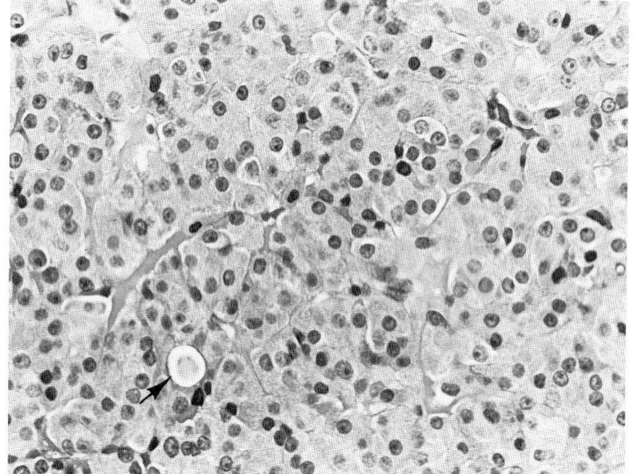

Figure 3.25. (A) Hürthle cell carcinoma infiltrating into the skeletal muscle adjacent to the thyroid gland (B). This Hürthle cell carcinoma presented as a periorbital metastatic tumor nodule. A central lumen (*arrow*) is present. The tumor cells were focally positive for thyroglobulin.

Variants of Follicular Carcinoma

Clear Cell Carcinoma

Rare follicular carcinomas (or adenomas) may be composed predominantly of clear cells. The appearance of the cells may be related to accumulation of glycogen, distended mitochondria, thyroglobulin, or other cytoplasmic vacuoles (15,69) (Fig. 3.23). These tumors behave like other follicular neoplasms, and capsular and vascular invasion are used to make the diagnosis of malignancy. Metastatic clear cell carcinomas to the thyroid as in renal cell carcinoma may look like clear cell thyroid carcinoma. Thyroglobulin staining, which is positive in the thyroid lesions, is helpful in the differential diagnosis.

Hürthle Cell Carcinoma

Some follicular carcinomas have granular eosinophilic cytoplasm due to the presence of abundant cytoplasmic mitochondria (Fig. 3.24, Fig. 3.25A and B). These tumors are known as Hürthle cell, oxyphilic, or eosinophilic cell carcinomas. Staining for thyroglobulin is focally positive in most tumors. Although the behavior of Hürthle cell tumor has been a controversial subject in the past (68), most of these neoplasms are benign and the same criteria that are used to diagnose other follicular neoplasms should be used to make the diagnosis of a Hürthle cell carcinoma (6,61). However, the proportion of malignant cases of Hürthle cell tumors may be higher than for other follicular neoplasms (61).

A

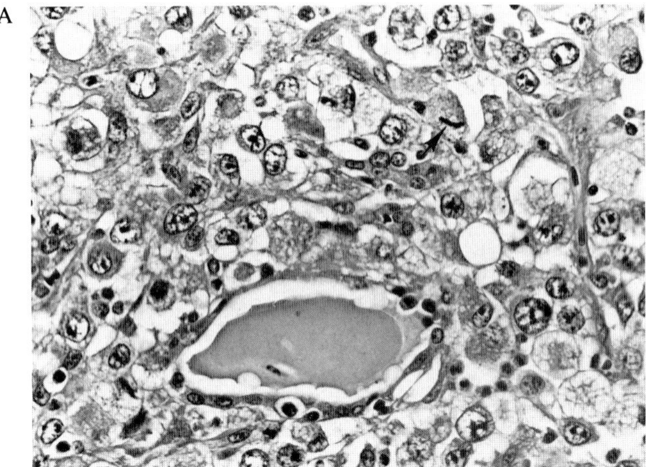

B

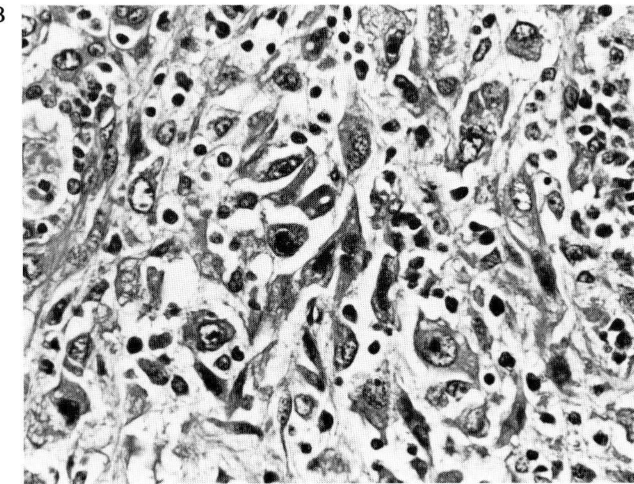

Figure 3.32. Anaplastic thyroid carcinoma (A) the tumor cells have pleomorphic nuclei and abundant cytoplasm. A mitotic figure (*arrow*) and an entrapped thyroid follicle and (B) marked cellular pleomorphism are present.

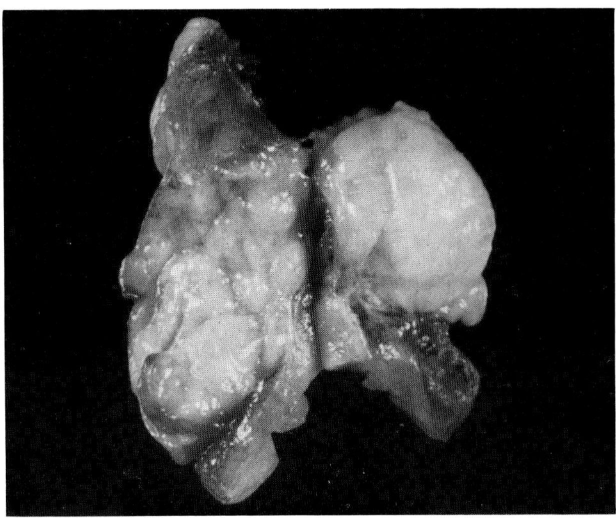

Figure 3.33. Malignant lymphoma of the thyroid. Most of the thyroid lobe has been replaced by the tumor cells. A residual portion of thyroid that contained lymphocytic thyroiditis is present in the lower part of the field.

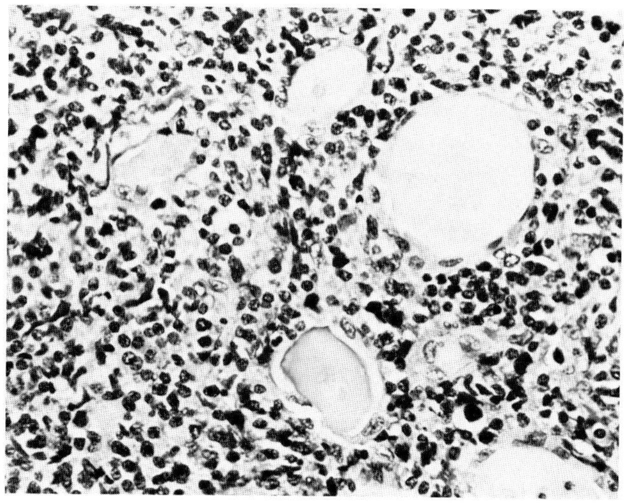

Figure 3.34. Malignant lymphoma. This small-cell lymphoma infiltrates around the residual follicles.

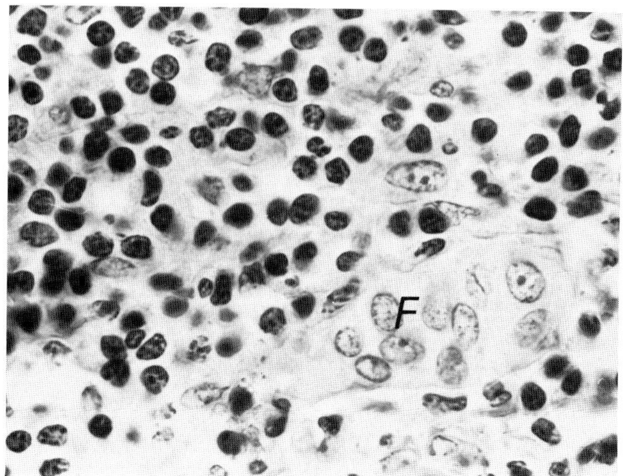

Figure 3.35. Another small-cell malignant lymphoma arising in a thyroid with lymphocytic thyroiditis. Residual follicular cells (*F*) are also seen.

with the subtype of lymphoma (Figs. 3.34, 3.35). Immunohistochemical stains on frozen and paraffin sections are helpful in confirming the diagnosis. Most thyroid lymphomas are of B-cell lineage.

Secondary lymphoma may involve the thyroid in patients with extensive disease.

Thyroid Sarcomas

These are rare malignant neoplasms. Well-documented angiosarcomas of the thyroid have been reported, primarily in Europe (20).

Treatment of Thyroid Malignancies

Most thyroid carcinomas are treated by excision, which may vary from partial lobectomy to total thyroidectomy depending on the subtype of tumor and degree of thyroid involvement (67). Anaplastic carcinomas are often treated by biopsy or excision with subsequent radiation and chemotherapy. With papillary and medullary carcinomas, which commonly involve cervical and other lymph nodes, individual lymph node resection is often done. Needle aspiration and frozen section examination are often done to establish the diagnosis of thyroid neoplasms. Needle aspiration is helpful in distinguishing inflammatory and neoplastic diseases

(52). It is often difficult to distinguish between benign and malignant follicular neoplasms by needle aspiration biopsies (53).

Radioactive iodine is very helpful in the treatment of follicular carcinomas (48). Hürthle cell tumors, as well as some papillary and anaplastic carcinomas, do not respond well to radioactive iodine. The general approach with radioactive iodine is to give a tracer dose to localize metastatic disease by scintiscan. A larger therapeutic dose is used to ablate metastatic disease. If a total thyroidectomy was not done, elimination of the residual thyroid is often necessary to get the optimal amounts of radioactive iodine to the tumor.

Chemotherapy and external beam irradiation are used for the more aggressive carcinomas, such as anaplastic carcinomas and recurrent carcinomas that have failed to respond to other therapeutic modalities (40). Lymphomas are often very responsive to external beam radiation and chemotherapy.

Prognosis

The prognosis of papillary carcinoma varies with the age of the patient. Patients less than 40 years of age and children have a much better prognosis. Direct extension of the tumor outside the thyroid is often associated with a worse prognosis. Occult or minimal papillary carcinomas are associated with an excellent prognosis, while columnar cell and tall-cell papillary carcinomas are associated with a poor prognosis.

Follicular carcinomas generally have a worse prognosis than papillary carcinomas. The size and the degree of invasiveness are related to the prognosis. Older patients usually have more invasive tumors. In minimally invasive carcinoma a 10-year survival of 86% has been reported, while in more invasive tumors the 10-year survival was 44% (77).

Familial medullary thyroid carcinoma (MEN 2a) usually has a better prognosis than sporadic tumors (14). A recent report by Samaan and co-workers suggests that the poorer prognosis associated with sporadic medullary thyroid carcinoma may be related to the patient's older age at the time of diagnosis rather than the inherent differences in the two forms of the disease (64). Additional studies will be needed to resolve these differences. Patients with MEN 2b have very aggressive medullary thyroid carcinomas.

Poorly differentiated or insular carcinomas have a worse prognosis than most well-differentiated carcinomas. In a study of 25 cases by Carcangui and her colleagues, 21 patients developed recurrent or metastatic disease and 14 patients died of their disease 1 to 8 years after initial therapy (9).

Anaplastic carcinoma is highly malignant. Many patients die of their disease several months after the initial diagnosis, although rare long-term survivors have been reported with very aggressive therapy (40).

References

1. Albores-Saavedra J, LiVolsi VA, Williams ED (1985) Medullary carcinoma. Semin Diag Pathol 2:137- 146
2. Batsakis JG, Nishiyama RH, Schmidt RW (1963) Sporadic goiter syndrome. A clinicopathologic analysis. Am J Clin Pathol 39:241–251
3. Bilano RA, Ward WQ, Little WP (1983) Minocycline and black thyroid. JAMA 249:1887
4. Block MA, Jackson CE, Greenwald KA, Yott JB, Tashjian AH Jr (1980) Clinical characteristics distinguishing hereditary from sporadic medullary thyroid carcinoma. Treatment implications. Arch Surg 115:142–148
5. Bocker W, Dralle H, Koch G, De Heer K, Hagemann J (1978) Immunohistochemical and electron microscope analysis of adenomas of the thyroid gland. II. Adenomas with specific cytological differentiation. Virchows Arch [Pathol Anat] 380: 205–220
6. Bondeson L, Bondeson A-G, Ljungberg O (1981) Oxyphil tumors of the thyroid. Ann Surg 194: 677–680
7. Campagno J, Oertel JE (1980) Malignant lymphoma and other lymphoproliferative disorders of the thyroid gland. A clinicopathologic study of 245 cases. Am J Clin Pathol 74:1–11
8. Carcangui ML, Steeper T, Zampi G, Rosai J (1985) Anaplastic thyroid carcinoma. A study of 70 cases. Am J Clin Pathol 83:135–158
9. Carcangui ML, Zampi G, Rosai J (1984) Poorly differentiated ("insular") thyroid carcinoma. A reinterpretation of Langhans "wuchernde Struma." Am J Surg Pathol 8:655–668
10. Carney JA, Moore SB, Northcutt RC, Woolner LB, Stillwell GK (1975) Palpation thyroiditis (multifocal granulomatous folliculitis). Am J Clin Pathol 64:639–647
11. Carney JA, Ryan J, Goellner JR (1987) Hyalinizing

trabecular adenoma of the thyroid gland. Am J Surg Pathol 11:583-591

12. Carney JA, Sizemore GW, Hayles AB (1978) Multiple endocrine neoplasia, type 2b. Pathol Biol 8: 105-153
13. Chen KTK, Rosai J (1977) Follicular variant of thyroid papillary carcinoma. A clinicopathologic study of six cases. Am J Surg Pathol 1:123-130
14. Chong GC, Beahrs OH, Sizemore GW, Woolner LH (1975) Medullary carcinoma of the thyroid gland. Cancer 35:695-699
15. Civantos F, Nadji M, Albores-Saavedra J, Morales AR (1984) Clear cell variant of thyroid carcinoma. Am J Surg Pathol 8:187-192
16. Cole WH, Slaughter DP, Majarkis JD (1949) Carcinoma of the thyroid gland. Surg Gynecol Obstet 89:349-356
17. Crile G Jr, Hazard JB (1953) Relationship of the age of the patient to the natural history and prognosis of carcinoma of the thyroid. Ann Surg 138: 33-38
18. DeLellis RA, Nunnemacher G, Wolfe HJ (1977) C-cell hyperplasia: An ultrastructural analysis. Lab Invest 36:237-248
19. DeLellis R, Wolfe HJ (1981) The pathobiology of the human calcitonin (C)-cell: A review. Pathol Annu 16:25-52
20. Egloff B (1983) The hemangioendothelioma of the thyroid. Virchows Arch [Pathol Anat] 400:119-142

20a. Elman DS (1958) Familial association of nerve deafness with nodular goiter and thyroid carcinoma. N Engl J Med 259:219-223

21. Evans HL (1986) Columnar cell carcinoma of the thyroid. A report of two cases of an aggressive variant of thyroid carcinoma. Am J Clin Pathol 85: 77-80
22. Evans HL (1987) Encapsulated papillary neoplasms of the thyroid. A study of 14 cases followed for a minimum of 10 years. Am J Surg Pathol 11:592-597
23. Fransilla KO, Ackerman LV, Brown CL, Hedinger CE (1985) Follicular carcinoma. Semin Diag Pathol 2:101-122
24. Fransilla KO (1975) Prognosis in thyroid carcinoma. Cancer 36:1138-1146
25. Fransilla KO, Harach HR, Wasenuis V-M (1984) Mucoepidermoid carcinoma of the thyroid. Histopathology 8:847-860
26. Frauenhoffer CM, Patchefsky AS, Cobanoglu A (1979) Thyroid carcinoma. A clinical and pathologic study of 125 cases. Cancer 43:2414-2421
27. Ginsberg J, Walfish P (1977) Post-partum transient thyrotoxicosis with painless thyroiditis. Lancet I: 1125-1126
28. Gross RE, Connerle ML (1940) Thyroglossal cysts and sinuses: a study and report of 198 cases. Engl J Med 223:616-624
29. Hanford JM, Quimby EH, Frantz VK (1962) Cancer arising many years after radiation therapy. Incidence after irradiation of benign lesions in the neck. JAMA 181:404-410
30. Hawk WA, Hazard JB (1976) The many appearánces of papillary carcinoma of the thyroid. Cleve Clin Q 43:207-216
31. Hazard JB, Hawk WA, Crile G (1959) Medullary (solid) carcinoma of the thyroid. A clinicopathologic entity. J Clin Endocrinol Metab 19:152-161
32. Hazard JB, Kenyon R (1954) Atypical adenoma of the thyroid. Arch Pathol 58:554-563
33. Hempelmann LH, Piter JW, Burke GF, Terry R, Ames WR (1967) Neoplasm in persons treated with x-rays in infancy for thymic enlargement. A report of the third follow-up survey. J Natl Cancer Inst 38:317-341
34. Horn RC (1951) Carcinoma of the thyroid. Description of a distinctive morphologic variant and report of seven cases. Cancer 4:697-707
35. Hubert JP, Kiernan PH, Beahrs OH, McConahey WM, Woolner LB (1980) Occult papillary carcinoma of the thyroid. Arch Surg 115:394-398
36. Ingbar SH (1985) The thyroid gland. In: Wilson JD, Foster DW (eds) Williams Textbook of Endocrinology. Philadelphia, Saunders, pp 682-815
37. Johnson TL, Lloyd RV, Thompson NW, Beierwaltes WH, Sisson JC (1988) Prognostic implications of the tall cell variant of papillary thyroid carcinoma. Am J Surg Pathol 12:22-27
38. Katz SM, Vickery AL Jr (1974) The fibrous variant of Hashimoto's thyroiditis. Hum Pathol 5:161-170
39. Kennedy JS, Thompson JA (1974) The changes in the thyroid gland after irradiation with 131-I or partial thyroidectomy for thyrotoxicosis. J Pathol 112:65-81
40. Kim JH, Leeper RD (1983) Treatment of anaplastic giant and spindle cell carcinoma of the thyroid gland with combination adriamycin and radiation therapy. A new approach. Cancer 52:954-957
41. Komorowski RA, Hanson GA (1977) Morphologic changes in the thyroid following low-dose childhood radiation. Arch Pathol Lab Med 101:36-39
42. Lang W, Georgii A, Stauch G, Kienzle E (1980) The differentiation of atypical adenomas and encapsulated follicular carcinomas in the thyroid gland. Virchows Arch [Pathol Anat] 385:125-141
43. Lee JG (1935) Chronic nonspecific thyroiditis. Arch Surg 31:982-1012
44. Lindsay S (1960) Carcinoma of the Thyroid Gland. Springfield, IL, Charles C. Thomas, p 43

CHAPTER 4
Parathyroid Glands

Development and Anatomy

The parathyroid glands arise from branchial pouches III and IV. They can first be recognized in the 8- to 9-mm embryo as bilateral proliferations along the pouches. During separation, fragmentation may occur and these may give rise to accessory parathyroid glands (parathyromatosis). The canals of Karsteiner are small epithelial tubules that may arise in association with parathyroids III and may persist into postnatal life (9). Parathyroid III migrates with the thymus, which also arises from the third pouches and comes to rest at the inferior poles of the thyroid gland and the inferior parathyroid. One or both parathyroid III's may remain attached to the thymus and go to the lower neck in the anterior mediastinum or in the posterior mediastinum (9). Parathyroid IV migrates only a short distance and becomes the superior pair of parathyroids near the superior poles of the thyroid gland. In rare cases either pair of parathyroid glands may remain in the pharyngeal wall. Supernumerary parathyroid glands have been reported in 2 to 65% of adults (9,15).

Each parathyroid gland weighs between 20 and 35 mg. The combined weight of the glands in men is 120 $\pm$ 3.5 mg and in women 142 $\pm$ 5.2 mg (14,21). Each gland is approximately 3 to 6 mm in length, 2 to 4 mm in width, and 0.5 to 2 mm in thickness (9). The glands are smooth and yellow to tan, depending on the amount of adipose tissue and oxyphil cells. The amount of interstitial adipose tissue increases with age. The superior parathyroids are supplied by small branches of the superior thyroid artery, while the inferior parathyroids are supplied by branches of the inferior thyroid artery. However, the vascular supply can be quite variable, especially with glands in aberrant locations.

Two principal types of cells are present in the glands. Most of the cells are chief cells, which have eosinophilic to amphophilic cytoplasm (Fig. 4.1). The oxyphil cells are larger than the chief cells and have eosinophilic cytoplasm. The number of these cells increases with age. The previous designation "water-clear cell" should probably be used only when the cells have membrane-limited vacuoles as seen in clear cell hyperplasia (9). The stromal fat cells increase with age and are uncommon before puberty. Stromal fat in the normal glands of adults ranges from 10 to 30% of the glandular weight (9,10,15).

Ultrastructural studies show that the chief cells undergo a cyclic process of parathyroid hormone synthesis and secretion (27). Dense-core secretory granules in the cytoplasm are variable but are generally few in number. Abundant intracellular fat is seen in the resting phase, especially in the presence of hypercalcemia. The intracellular fat can be identified by light microscopy with lipid and other stains such as Oil red O, Sudan V and azure A, and Erie garnet B (6,23). The oxyphil cells have abundant mitochondria, which accounts for their bright eosinophilic cytoplasm. They usually do not contain significant amounts of lipid.

Abnormalities of the Parathyroids

Ectopic Parathyroid

Ectopic glands are not uncommon, especially from parathyroid III along their migration path. In a

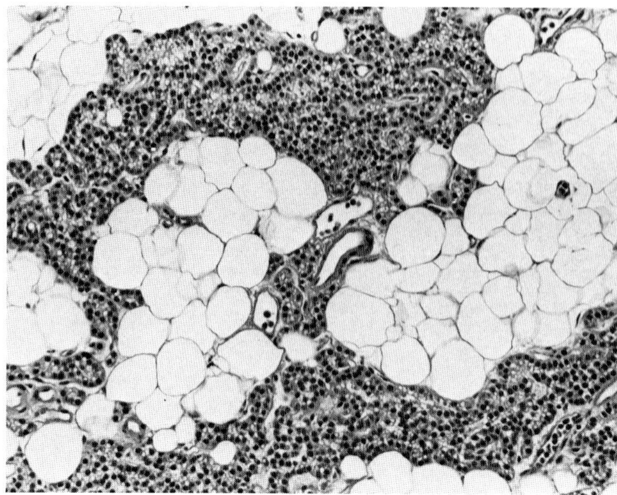

Figure 4.1. Normal parathyroid gland showing abundant stromal fat and a predominance of chief cells.

large series reported by Wang, 2% of parathyroids III were ectopic, 39% were in the thymic tongue, and 2% were in the mediastinal portion of the thymus, while 1% of parathyroids IV were retropharyngeal and retroesophageal (31).

Parathyroid Cysts

Primary cysts of the parathyroid may present as a neck mass. They usually do not have any endocrine function, although the cyst fluid may have significant amounts of parathyroid hormone (32). Most cysts are located in the lower portion of the neck and are often loosely attached to the thyroid gland. They can range from 2.5 to 25 cm in diameter. The wall is gray-white and translucent. Small clusters of parathyroid cells are often present in the wall. The lining may be acellular or have flattened chief cells. Parathyroid cysts may arise from degenerating adenomatous or hyperplastic glands.

Infarction

In rare cases surgical exploration of the parathyroids or trauma may compromise the blood supply and lead to infarction (33).

Hypoparathyroidism

This condition is characterized by hypocalcemia and neuromuscular symptoms. It can result from deficiency of parathyroid hormone production or end organ resistance to the action of parathyroid hormone.

Postoperative hypoparathyroidism may be caused by excision of the parathyroids or damage to the glands or to their vascular supply during surgery for hyperparathyroidism, thyroid disorder, or radical neck dissections for malignancies. The symptoms usually develop a few days after surgery, but they may develop after longer postsurgical intervals.

Idiopathic Hypoparathyroidism

This rare condition may develop a) as an isolated condition, b) in association with DiGeorge syndrome with agenesis of the thymus, or c) in association with a familial disorder involving the thyroid and adrenal glands.

The DiGeorge syndrome involves congenital absence of the thymus and parathyroid glands. Patients usually have persistent infection due to cellular immune deficiency and hypocalcemia, both of which can lead to death.

Familial hypoparathyroidism may have an autoimmune etiology. Patients may have antibodies against the parathyroids, adrenal cortex, thyroid, parietal cells, and ovaries in females (5).

Pseudohypoparathyroidism

This is an inherited disorder characterized by hypoparathyroidism with hypocalcemia and hyperphosphatemia associated with skeletal and developmental defects. There is deficient end organ response to endogenous parathyroid hormone leading to hypoparathyroidism. Patients usually have a deficient or absent increase in urinary cyclic adenosine monophosphate excretion in response to parathyroid hormone, along with hyperplasia of the parathyroid glands. The biochemical abnormality is in the parathyroid hormone receptor-adenyl cyclase complex. These patients usually have an increased concentration of circulating parathyroid hormone. Relatives often exhibit skeletal and developmental defects without any symptoms or chemical evidence of hypoparathyroidism—a condition known as pseudopseudohypoparathyroidism.

Hyperparathyroidism

Primary hyperparathyroidism is a common cause of hypercalcemia. It is usually caused by adenomas (81%), hyperplasia (16%), and carcinomas (3%). Primary hyperparathyroidism is associated with an increased secretion of parathyroid hormone that leads to hypercalcemia and hypophosphatemia. Other problems, including recurrent nephrolithiases, peptic ulcers, bone destruction, and mental changes, may also be present.

Primary hyperparathyroidism is unusual in infants and children. This disease is more common in older patients, with a mean age of 50 years (9). Patients in the second and third decades of life with primary hyperparathyroidism may have multiple endocrine neoplasia with parathyroid hyperplasias or adenomas. Secondary hyperparathyroidism is most commonly secondary to renal disease.

There is a slight female preponderance with adenomas and chief cell hyperplasia (3:2), while carcinoma is equally frequent in both sexes.

Primary Parathyroid Hyperplasia

These diseases include hyperplasia of chief cells and/or of clear cells. Primary chief cell hyperplasia is the most common type. About half the patients may have equal enlargement of all four glands, while the remaining half of the glands may be very enlarged or about normal in size (2,9). Parathyromatosis or supernumerary glands may become enlarged from stimulation in primary hyperplasia.

The gross appearance of the glands in chief cell hyperplasia varies from nodular to diffuse enlargement with a yellow-tan to red-brown appearance. Cystic change may be present. The glands are usually variably enlarged and some glands may appear to be of normal size (Fig. 4.2).

The microscopic appearance can include arrangement of the cells in cords, sheets, and follicules with reduction in the stromal fat (Fig. 4.3). A mixture of chief, oxyphil, and transitional oxyphil cells is commonly seen. Glycogen accumulation in the cytoplasm may give the cells a clear cell appearance. Nuclear pleomorphism is uncommon.

Occasionally the residual adipocytes in the glands may simulate a rim of normal parathyroid tissue, a condition that is seen more commonly, but not exclusively, with adenomas.

Fat stains at the time of frozen section show decreased intracytoplasmic fat (Fig. 4.4A and B). In infants and children there is little stromal fat, so the increased weight of the gland is one of the few helpful diagnostic features. Immunohistochemical staining for parathyroid hormone is positive but does not help to distinguish a normal from an abnormal gland.

Clear cell or "water-clear cell" hyperplasia is associated with enlarged irregular glands, which are seldom nodular. They are usually soft, brown, and sometimes cystic. Microscopic nodularity is not present. Cells with large clear cytoplasm and central nuclei that look like those in renal cell carcinoma are present. Ultrastructural examination reveals vacuoles, which are thought to be derived from the Golgi apparatus. Secretory granules can also be identified.

Neonatal hyperparathyroidism is a rare condition that can lead to severe hypercalcemia and death if not corrected (12). The combined parathyroids usually weigh 200 mg, while the normal weight of the four glands in neonates is 4 to 20 mg. Histological examination shows diffuse sheets and cords of vacuolated cells containing glycogen. Stromal fat is sparse in neonates, so its absence in this condition is not helpful diagnostically.

Adenomas

Although most adenomas develop in only one of the four glands, unusual cases of intrathyroidal adenomas, adenomas arising in a supernumerary gland, and adenomas arising in the mediastinum may occur. Most lesions diagnosed as double ade-

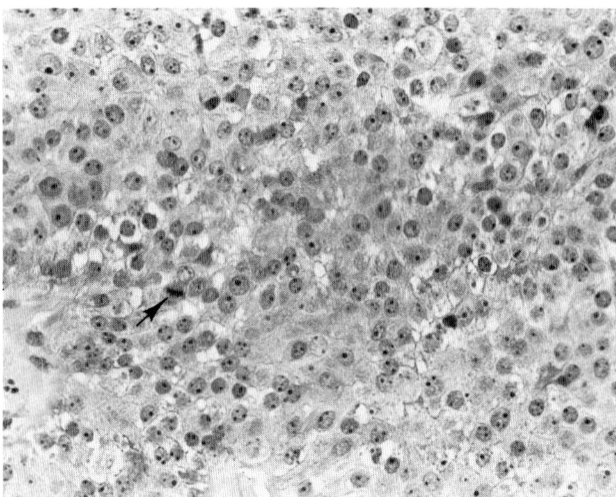

Figure 4.10. Parathyroid carcinoma. The tumor cells have large nuclei and several nucleoli. Increased mitotic activity is present (*arrow*).

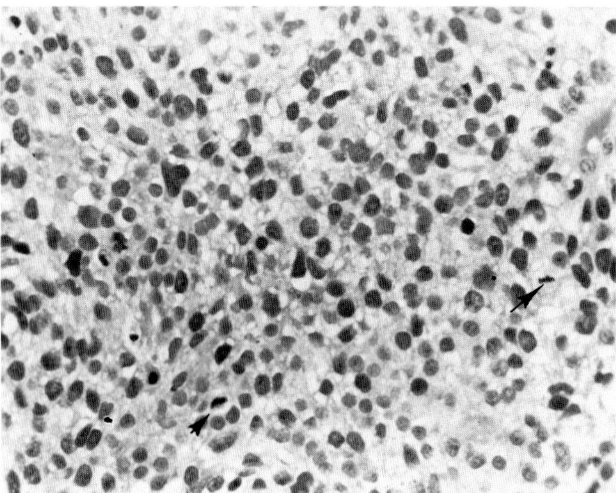

Figure 4.11. Parathyroid carcinoma. The tumor cells are relatively uniform and several mitotic figures (*arrows*) are present.

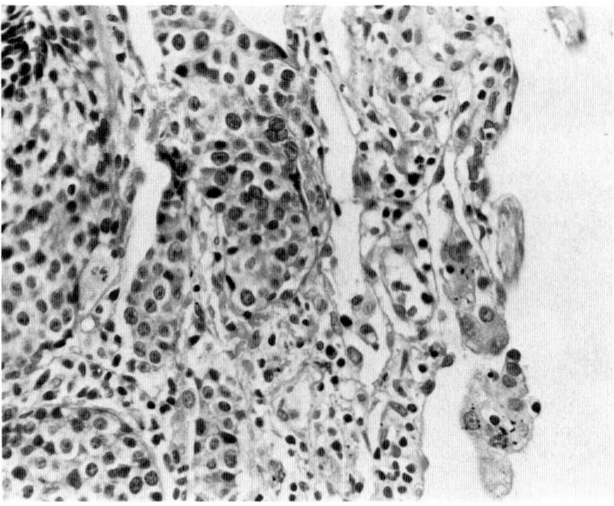

Figure 4.12. Parathyroid carcinoma metastatic to the lung. The tumor cells appear quite uniform. The tumor was initially diagnosed as a metastatic carcinoid. The patient was subsequently found to have a serum calcium of 21 mg per deciliter.

tissues. Carcinomas usually range in size from 1 to 6 cm.

Microscopically, carcinomas often have a prominent trabecular pattern with fibrous bands surrounding nests of tumor cells. Increased mitotic activity and blood vessel and capsular invasion are frequently seen (Figs. 4.10 to 4.12). Nuclear palisading is sometimes present. Cellular atypia and nuclear pleomorphism are not helpful features in distinguishing adenomas from carcinomas, since these features are common in adenomas. Parathyroid carcinomas often have uniform cells with minimal pleomorphism. The cells are often larger and more regular than those in adenomas.

Lipid stains show decreased intracellular lipid. Ultrastructural studies show chief cells in the active phase of secretion, which correlates with the usual high degree of hypercalcemia in these tumors.

Local recurrences and metastases to regional lymph nodes, lungs, liver, and bone are common (24,26). When death results from parathyroid carcinoma it is usually due to the complications of hypercalcemia rather than to widespread metastasis.

Nonfunctional parathyroid carcinomas are rare (18,20). These tumors may contain immunoreactive parathyroid hormone, but the serum level of parathyroid hormone is often within the normal range.

Secondary Hyperparathyroidism

This condition is associated with increased secretion of parathyroid hormone secondary to extraglandular stimulation of the parathyroid glands. The principal stimulus for increased secretion is depression of the serum levels of ionized calcium. Chronic calcium suppression leads to parathyroid gland hyperplasia. Renal disease is the most common cause of secondary hyperparathyroidism. Vitamin D deficiency and malabsorption may also be etiologic factors.

The gross appearance is similar to that of primary chief cell hyperplasia (Fig. 4.13). Microscopic evidence of decreased stromal lipid and intracellular lipid is often present. In many cases nodularity may be prominent and oxyphil cells are sometimes more prominent than in primary hyperplasia (Fig. 4.14).

Tertiary Hyperparathyroidism

This condition can develop in patients with a history of secondary hyperparathyroidism who have depressed calcium and elevated serum phosphate levels and then go on to develop elevated serum calcium. They often have severe renal disease or malabsorption or a history of renal transplantation (1). The glands become autonomous and do not respond to physiologic controls. Grossly, a dominant nodule may be present in one or more glands. The histologic picture is similar to that of primary and secondary hyperparathyroidism. A dominant nodule resembling an adenoma may be seen in the histologic sections (9).

Differential Diagnosis of Hyperparathyroidism

To distinguish between an abnormal and a normal parathyroid gland, the lipid stain to demonstrate intracellular lipid and the weight of the gland are most helpful. In most normal parathyroids there is abundant intracellular lipid, whereas in hyperplastic and neoplastic glands the intracellular lipid is decreased. In addition, the lipid droplets are usually larger in the normal gland than in adenomas or hyperplasias. About 10 to 15% of hyperplastic and adenomatous glands may have normal amounts of intracellular lipid but the lipid droplets are often smaller than those in the normal glands. Oxyphil cells also have little intracellular lipid even in normal glands. Stromal fat may vary, depending on the part of the gland that was sampled and the age of the patient, since children have little stromal fat.

It is necessary to examine at least two glands and preferably four to distinguish between adenomas and hyperplasias. In adenomas one gland is enlarged and the others are normal. Recent restriction fragment length polymorphism analyses enabled Arnold and colleagues to distinguish between parathyroid adenomas and hyperplasia with molecular biological techniques (4). Double adenomas are extremely rare and most reported cases probably represent early hyperplasia in the other glands (16). Hyperplasia can be quite variable, with two or more glands appearing almost normal in size with normal intracellular lipid content. Recent evidence from molecular biological studies revealed that ade-

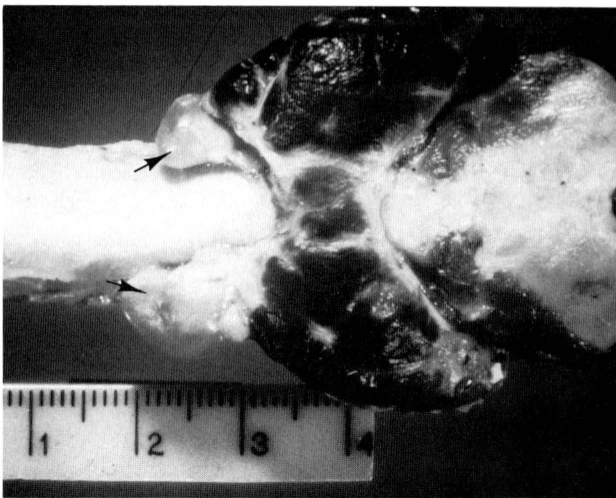

Figure 4.13. Secondary hyperparathyroidism. The patient had chronic renal failure. The two inferior parathyroids are markedly enlarged, with the left inferior gland being much larger than the right inferior gland (*arrows*).

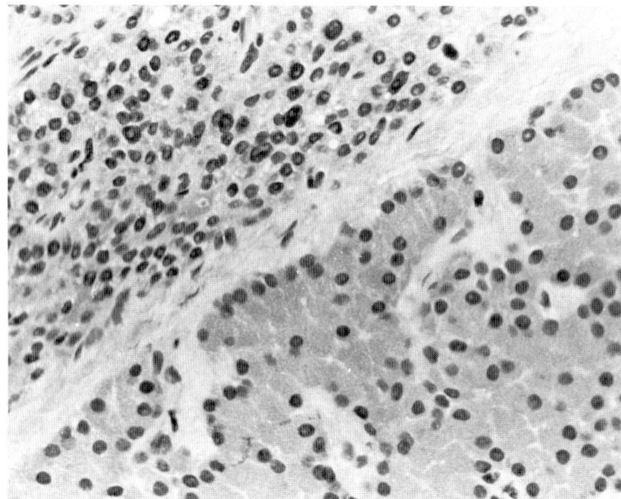

Figure 4.14. Parathyroid hyperplasia secondary to chronic renal failure. Prominent oncocytic cell nodules are present in the gland.

nomas are monoclonal while hyperplastic glands are polyclonal (4).

The differential diagnosis between adenoma and carcinoma relies on increased numbers of mitoses, fibrosis, and vascular invasion. The diagnosis can sometimes be difficult and a diagnosis of malignancy may have to be deferred if it cannot be made with certainty by histology in the absence of metastasis. Such rare borderline lesions should probably be designated as parathyroid neoplasms of uncertain biological potential. Recurrent adenomas from incomplete surgical removal may have extensive fibrosis secondary to previous surgery.

Chief cell hyperplasia may have the appearance of renal cell carcinoma, but this distinction can usually be made easily by examining several glands and from the clinical history. Metastatic carcinomas to the parathyroid gland are rare but can occur with carcinoma of breast, prostate, lung, kidney and others. Immunohistochemical staining for parathyroid hormone and other markers can distinguish a parathyroid lesion from other neuroendocrine carcinomas even if the parathyroid tumor is not functional.

Treatment

Surgery is the treatment of choice for most parathyroid lesions causing hyperparathyroidism. The number of glands that are removed for hyperplasia

Figure 4.15. Bone from a patient with hyperparathyroidism. There is increased osteoclastic activity (*arrows*) and extensive marrow fibrosis is present.

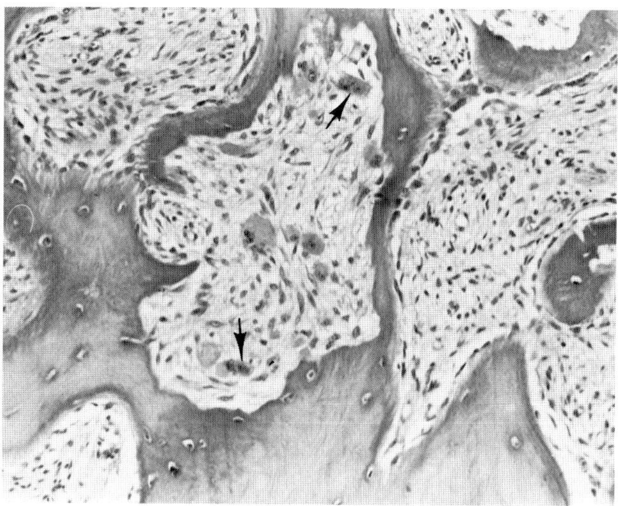

is variable (8,30). Biopsy of at least two glands and preferably four should be done. Surgical treatment of recurrent hyperparathyroidism when an ectopic gland is suspected may require extensive neck and mediastinal exploration. Some adenomas may also be intrathyroidal.

Parathyroid carcinoma is usually treated by removal of the tumor en bloc to prevent local recurrence. In some cases combined chemotherapy may be helpful (24).

A slightly increased incidence of parathyroid adenoma occurring up to 30 years after head and neck radiation has been reported (17).

Patients with multiple endocrine neoplasia types 1 and 2a often have nodular parathyroid hyperplasia, although adenomas have also been reported (3). The mean age of these patients with hypercalcemia is much younger than that of patients with sporadic hyperparathyroidism.

Prognosis

The results of surgical treatment for parathyroid hyperplasia and adenoma are variable (30). Persistent hypercalcemia may be related to ectopic parathyroids, mediastinal adenomas, and incorrect clinical and pathologic diagnoses such as in patients with sarcoidosis (30).

Parathyroid carcinomas are often associated with recurrent diseases, and metabolic effects of hypercalcemia have been a common cause of death. Less than 50% of patients die of this disease in 5 years (26).

Lesions Associated with Hyperparathyroidism

The major effects of parathyroid hormone on bone are to stimulate formation of osteoclasts (Fig. 4.15). Excessive secretion of parathyroid hormone results in osteitis fibrosa cystica or von Recklinghausen's disease of bone. This is characterized by resorption of bone by increased osteoclastic activity, replacement of marrow spaces and resorbed bone by fibrous tissue, and osteoblasts adjacent to newly deposited osteoid. The "brown tumor" of hyperparathyroidism, which is now rarely seen in the United States, is characterized histologically by proliferation of osteoclasts, hemorrhage, and macrophages with hemosiderin in a fibrous tissue background.

Other effects of hyperparathyroidism include increased formation of kidney stones and dystrophic calcium deposition in multiple sites including the skin.

Hypocalcemia Associated with Nonparathyroid Malignancies

Patients commonly have hypercalcemia and hypophosphatemia associated with malignancy such as squamous cell carcinoma of lung and renal cell carcinoma. Some tumors secrete a parathyroid hormone-like substance. There is evidence that these tumors are *not* producing parathyroid hormone messenger RNA (28) but in many cases produce parathyroid hormone-related peptide (7). (See discussion in Chapter 11.) The parathyroid glands in patients with hypercalcemia associated with other neoplasms often have normal weights and normal cell-to-fat ratios (11).

References

1. Abul-Haj SK, Conklin H, Hewitt WC (1962) Functioning lipoadenoma of the parathyroid gland. Report of a unique case. N Engl J Med 266:121–123
2. Akerstrom G, Rudberg C, Grimelius L, Bergstrom R, Johansson H, Ljunghall S, Rastad J (1986) Histologic parathyroid abnormalities in an autopsy series. Hum Pathol 17:520–527
3. Allo M, Thompson NW (1982) Familial hyperparathyroidism caused by solitary adenomas. Surgery 92:486–490
4. Arnold A, Staunton CE, Kim HG, Randall AB, Gaz D, Kronenberg HM (1988) Monoclonality and abnormal parathyroid hormone genes in parathyroid adenomas. N Engl J Med 318:658–662
5. Blizzard RM, Chee D, Davis W (1966) The incidence of parathyroid and other antibodies in the sera of patients with idiopathic hypoparathyroidism. Clin Exp Immunol 1:119–128
6. Bondeson A-G, Bondeson L, Ljungberg O, Tibblin S (1985) Fat staining in parathyroid disease. Diagnostic value and impact on surgical strategy: Clinicopathologic analysis of 191 cases. Hum Pathol 16:1255–1263
7. Broadus AE, Mangin M, Ikeda K, Insoga KL, Weir EC, Burtis WJ, Stewart AF (1988) Humoral hypercalcemia of cancer. Identification of a novel parathyroid hormone-like peptide. N Engl J Med 319: 556–563
8. Bruining H (1971) Surgical Treatment of Hyperparathyroidism. Springfield, IL, Charles C Thomas
9. Castleman B, Roth SI (1978) Tumors of the parathyroid glands. In: Atlas of Tumor Pathology, Second Series, Fascicle 14. Washington, DC, Armed Forces Institute of Pathology
10. Dekker A, Dunsford HA, Geyer SJ (1979) The normal parathyroid gland at autopsy: The significance of stromal fat in adult patients. J Pathol 128:127–132
11. Dufour DR, Marx SJ, Spiegel AM (1985) Parathyroid gland morphology in nonparathyroid hormone-mediated hypercalcemia. Am J Surg Pathol 9:43–51
12. Barcia-Bunuel R, Kutchemeshgi A, Brandes D (1974) Hereditary hyperparathyroidism. The fine structure of the parathyroid gland. Arch Pathol 97: 399–403
13. Geelhoed GW (1982) Parathyroid adenolipoma: clinical and morphologic features. Surgery 92: 806–810
14. Gilmour JR, Martin WJ (1937) The weight of the parathyroid glands. J Pathol Bacteriol 44:431–462
15. Grimelius L, Akerstrom G, Johansson H, Bergstrom R (1981) Anatomy and histopathology of human parathyroid glands. Pathol Annu 16:1–24
16. Harness JK, Ramsbury SR, Nishiyama RH, Thompason NW (1979) Multiple adenomas of the parathyroids: do they exist? Arch Surg 114:468–474
17. Hedman I, Tisell L-E (1983) Associated hyperparathyroidism and nonmedullary thyroid carcinoma: The etiologic role of radiation. Surgery 95: 392–397
18. Murphy MN, Glennon PG, Diocee MS, Wick MR, Cavers DJ (1986) Nonsecretory parathyroid carcinoma of the mediastinum. Light microscopic, immunocytochemical and ultrastructural features of a case and review of the literature. Cancer 58: 2468–2476
19. Ordonez NG, Ibanez ML, Mackay B, Samaan NA, Hickey RC (1982) Functioning oxyphil cell adenomas of parathyroid glands: Immunoperoxidase evidence of hormonal activity in oxyphil cells. Am J Clin Pathol 78:681–689
20. Ordonez NG, Ibanez ML, Samaan NA, Hickey RC (1983) Immunoperoxidase study of uncommon parathyroid tumors. Am J Surg Pathol 7:535–542
21. Roth SI (1971) Recent advances in parathyroid gland pathology. Am J Med 50:612–622
22. Roth SI, Gallagher MJ (1976) The rapid identification of "normal" parathyroid glands by the presence of intracellular fat. Am J Pathol 84:521–528
23. Saffos RO, Rhatigan RM (1979) Intracellular lipid in parathyroid glands. Hum Pathol 10:483–485
24. Schantz A, Castleman B (1973) Parathyroid carcinoma: A study of 70 cases. Cancer 31:600–605
25. Selzman HM, Fechner RE (1967) Oxyphil adenoma and primary hyperparathyroidism. Clinical and ultrastructural observations. JAMA 199:359–361
26. Shane E, Bilezikian JP (1982) Parathyroid carcinoma: A review of 62 patients. Endocrine Rev 3: 218–226
27. Shannon WA, Roth SI (1974) An ultrastructural

study of acid phosphatase activity in normal adenomatous and hyperplastic (chief-cell type) human parathyroid glands. Am J Pathol 77:493–506

28. Simpson EL, Mundy GR, D'Souza SM, Ibbotson KJ, Bockman R, Jacobs JW (1983) Absence of parathyroid hormone messenger RNA in nonparathyroid tumors associated with hypercalcemia. N Engl J Med 309:325–330
29. Snover DC, Foucar K (1981) Mitotic activity in benign parathyroid disease. Am J Clin Pathol 75: 345–347

29a. Stork PJ, Herteaux C, Frazier R, Kronenburg H, Wolfe HJ (1989) Expression and distribution of parathyroid hormone and parathyroid hormone messenger RNA in pathological conditions of parathyroid. Lab Invest 60:92A (abstract 547)

30. Thompson NW, Eckhauser FE, Harness JK (1982) The anatomy of primary hyperparathyroidism. Surgery 92:814–821
31. Wang C-A (1976) The anatomic basis of parathyroid surgery. Ann Surg 183:271–275
32. Wang C-A, Vickery AL Jr, Maloof F (1972) Large parathyroid cysts mimicking thyroid nodules. Ann Surg 175:448–453
33. Warner NE (1971) Basic Endocrine Pathology. Chicago, Year Book Medical Publishers
34. Wolpert HR, Vickery AL, Wang C-A (1989) Functioning oxyphil cell adenomas of the parathyroid gland. A study of 15 cases. Am J Surg Pathol 13: 500–504

CHAPTER 5

Endocrine Pancreas

Development and Anatomy

The pancreas arises from a ventral and a dorsal bud, which fuse to form the definitive pancreas in the 6-week fetus. The gland, which is derived from the primitive duodenum, is of endodermal origin. The ducts and acini arise from secondary branchings of the buds, while the islets of Langerhans, which also arise as epithelial outbranchings, lose their connections to the ductal lumina early in fetal life and can be identified in the 7-week fetus. These outbranchings give rise to isolated nests, which later become the definitive islets (50a,50b). Thus, the islet cells, like the exocrine pancreas, are of endodermal origin.

The adult pancreas has 300,000 to 2,000,000 islets. The normal islets have a maximum diameter of about 300 μm. The average size of an islet is approximately 150 μm. The islets are scattered throughout the exocrine pancreas, but they are thought to be more numerous in the tail of the pancreas than in the body or head of the gland.

The islets receive their blood supply from branches of the interlobular arteries that arise from the celiac trunk and the superior mesenteric artery. Unmyelinated nerve fibers penetrate into the islets but their function is not completely understood.

Various cell types can be identified in the normal islets (63). Neuron-specific enolase and chromogranin A messenger RNA are expressed by all islet cells (Figs. 5.1, 5.2). The alpha or glucagon-producing cells constitute 15 to 20% of the islet cells (Fig. 5.3, Table 5.1). They stain very strongly with argyrophil stains and with chromogranin A antibodies and are located predominantly at the periphery of the islets. The ultrastructural features include secretory granules 250 nm in diameter with a dense central core and a peripheral clear space. The beta or insulin-producing cells constitute about 60 to 80% of islet cells (Fig. 5.4). They are centrally located and stain positive for aldehyde fuchsin but weakly with chromogranin A antibodies. The ultrastructural features of the beta cells include irregular crystalline or eccentric dense-core granules separated from the membrane.

The delta or somatostatin-producing cells are usually interspersed with alpha or beta cells. The Hellerstrom-Hellman argyrophilic technique stains these cells strongly. They constitute less than 5% of islet cells (Fig. 5.5). The secretory granules range from 150 to 350 nm in diameter (Fig. 5.6). The pancreatic polypeptide or F cells constitute less than 2% of islet cells (Fig. 5.6). Pancreatic polypeptide-producing cells (Fig. 5.7) are also present in small and medium-size pancreatic ducts and with acinar cells. These cells have small secretory granules 140 to 200 nm in diameter.

The EC or enterochromaffin cells are present in variable numbers in the islets. They are argyrophilic and the ultrastructure is similar to that of the EC cells in the gastrointestinal tract.

Gastrin-producing cells are not present in the adult human pancreas but are present during fetal life. Likewise, vasoactive intestinal polypeptide-producing cells are not present in the adult pancreas.

Various drugs can damage the islet cells. Streptozotocin (9) and alloxan cause selective necrosis of beta cells and cobaltous chloride causes reversible damage to the alpha cells, including vacuolation and degranulation.

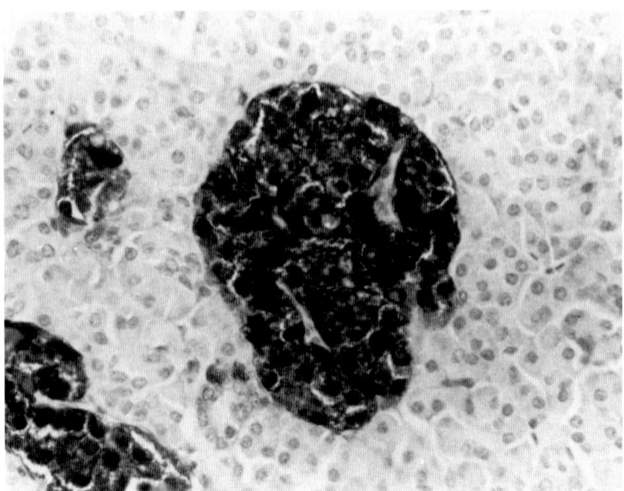

Figure 5.1. Immunostaining for neuron-specific enolase in pancreatic islets. All of the islet cells express this protein.

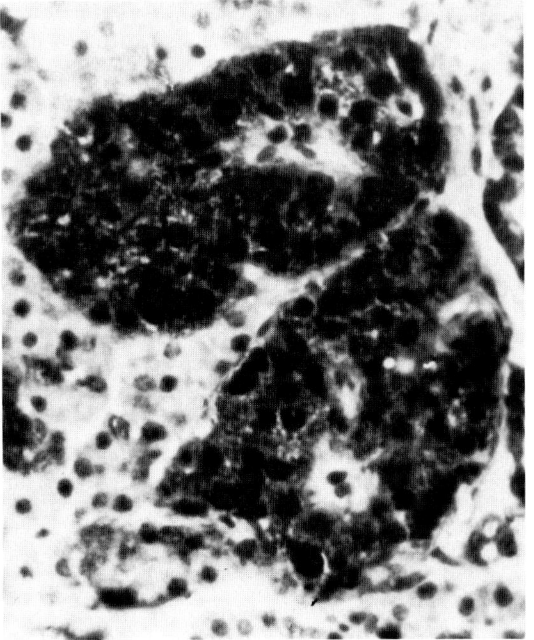

Figure 5.2. In situ hybridization to localize chromogranin A in islet cells with biotinylated oligonucleotide probe. All of the islet cells express chromogranin A messenger ribonucleic acid.

Diabetes Mellitus

This is a heterogeneous group of disorders associated with hyperglycemia due to a relative or absolute deficiency of insulin in the presence of a relative or absolute excess of glucagon (81). It is usually defined by a fasting plasma glucose level greater than or equal to 140 mg/dl on at least two occasions or an oral glucose tolerance test with a plasma glucose level greater than or equal to 200 mg/dl at 2 hours and one other point in the test. Diabetes is associated with several late complications including diseases affecting the blood vessels, eyes, kidney, and nerves. It is a major cause of renal failure, myocardial infarction, stroke, and blindness. Diabetes mellitus is divided into two major categories. Type I diabetes mellitus (insulin-dependent diabetes mellitus, IDDM) and type II (non-insulin-dependent diabetes mellitus, NIDDM) (Table 5.2). Patients with type I DM usually have absent or decreased beta cell function and insulin secretion, and exogenous insulin is usually required to prevent diabetic ketoacidosis. In type II DM, patients have normal or almost normal beta cell function.

Type I or Insulin-Dependent Diabetes Mellitus (IDDM)

The prevalence of type I diabetes in the United States is about 260/100,000 by age 20 (47) and only a few patients develop type I diabetes after age 20. The incidence range in the United States for patients with type I diabetes is between 10 and 16 per 100,000 (47). The etiologic factors associated with type I DM include genetic factors, viruses, and immunologic factors.

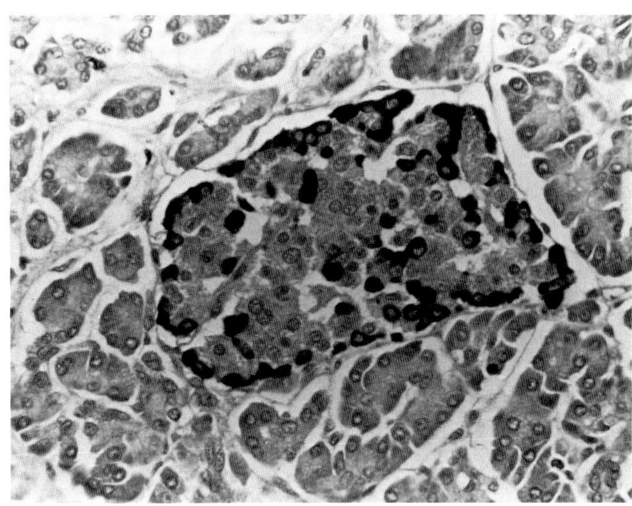

Figure 5.3. Immunostaining with a glucagon antibody revealing glucagon-producing cells predominantly around the periphery of the islet.

Table 5.1. Features of pancreatic endocrine cells

Cell type	Percent of cells	LM stain	Granule ultrastructure	Immunochemistry
Glucagon (α)	15–20	Rhodacyan Grimelius (argyrophil)	Targetoid dense core 250 nm	Chromogranin A, strongly positive NSE
Insulin (β)	60–80	Aldehyde thionine Aldehyde fuchsin	Irregular crystalline 100–300 nm	Chromogranin A, weakly positive NSE, synaptophysin
Somatostatin (δ)	5	Hellerstrom-Hellman	Regular 150–350 nm	NSE, synaptophysin
Pancreatic polypeptide (F)	2	Grimelius (argyrophil)	Regular 140–200 nm	Chromogranin A and NSE positive
Enterochromaffin (EC)	<1	Grimelius (argyrophil)	Pleomorphic	Serotonin positive

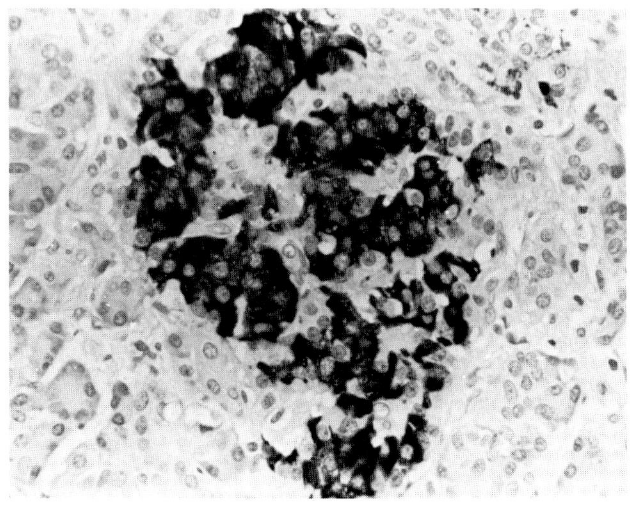

Figure 5.4. Insulin-producing cells constitute 60 to 80% of islet cells.

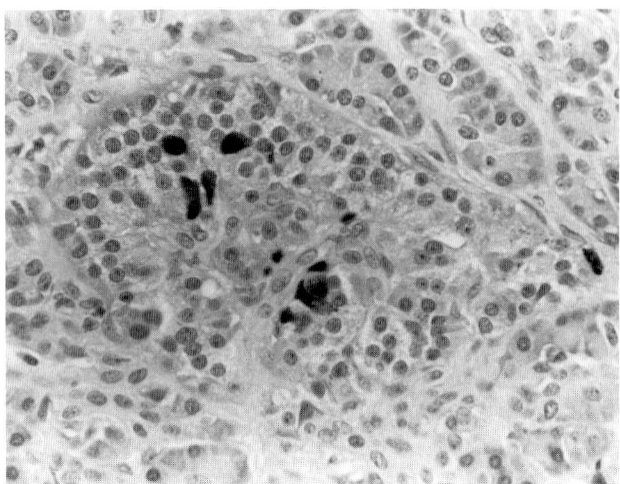

Figure 5.5. Somatostatin-producing cells in the islets. These cells have elongated cytoplasmic processes that may contact other islet cells. They have a paracrine regulatory function.

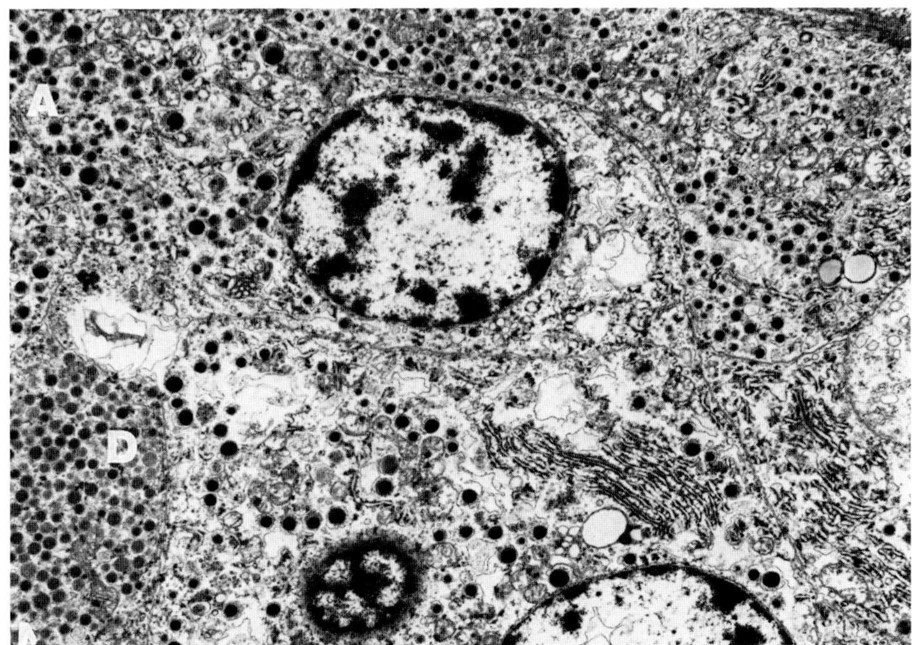

Figure 5.6. Electron micrograph of islet cells revealing a somatostatin-producing cell (*D*) and many glucagon-producing cells with a dense central core (*A*) (×6450).

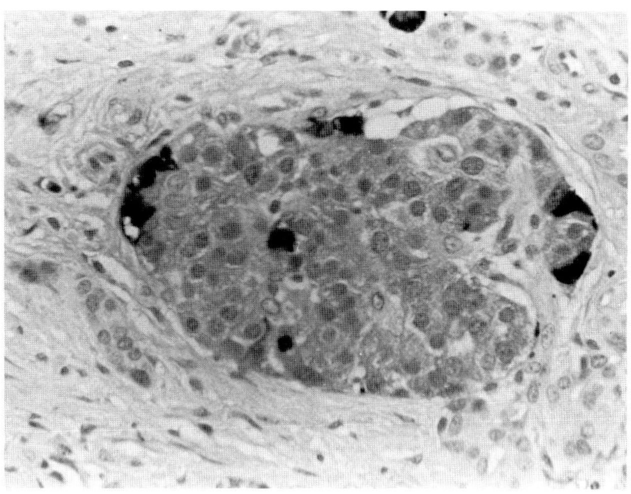

Figure 5.7. Pancreatic polypeptide-producing cells constitute about 2% of islet cells but increase in number in chronic pancreatitis.

Table 5.2. Features of insulin-dependent and non-insulin-dependent diabetes mellitus

Feature	Type I	Type II
Age of onset	Usually before 20 years	After 20 years
Body weight	Normal	Usually increased
Role of genetics	Small	Significant
Monozygotic twins	<50% concordant	<90% concordant
HLA association	Common	Rare
Cytoplasmic islet cell antibodies	Positive	Negative
Beta cells	Markedly decreased	Normal to slightly decreased
Insulin and C peptide in plasma	Absent or low	High

Genetics

Susceptibility to type I DM is linked to specific HLA alleles. Both class I and class II antigens are associated with the development of type I DM. Many patients with type I DM have either DDR3 or DDR4 antigens (74,84). A significant percentage of these patients have both DR3 and DR4 antigens. The mode of inheritance of type I DM is unknown. The current thought is that there are at least two susceptibility genes which act synergistically (81). One gene would put the subject at risk for diabetes and the second gene would enhance the risk.

Role of Viruses in the Etiology of IDDM

The observation that the onset of IDDM occurred with or shortly after infection with specific viruses such as mumps, rubella, measles, influenza, or Epstein-Barr virus strongly suggested a possible etiologic role of viruses (61). Menser and coworkers (57) noted that children with neonatal rubella have an increased risk of developing type I DM. Coxsackie B4 virus has been strongly implicated as an etiologic agent of type I DM (87). This virus was isolated from a patient who died with postinfectious diabetes and ketoacidosis. The virus that was isolated from the pancreas caused hyperglycemia when injected into mice (87). Most of the observations linking viruses with DM are indirect, but the evidence is very suggestive of a role of viruses in the pathogenesis of IDDM.

Immunologic Factor in IDDM

Several lines of evidence suggest that IDDM may have an immune-mediated etiology. Patients with recent diagnoses of type I diabetes mellitus usually have serum antibodies that react with islet cells (60 to 90%) of cases, while these antibodies are rare in individuals without diabetes (50). Interestingly, these antibodies react with all of the major islet cells, although only the beta cells are selectively destroyed in type I DM. Islet cell cytoplasmic antibodies are found in 85 to 90% of patients with type I DM up to 2 years after the onset of their disease (33).

Activated T lymphocytes, including helper and cytotoxic cells, are often associated with type I DM. This suggests that both cellular and humoral mediated immunity may be involved (20,30).

Type II or Non-Insulin-Dependent Diabetes Mellitus

This form of diabetes is more common than type I (1). There is a high prevalence among some groups in the United States such as the Pima indians of Arizona, where more than 40% of the population has the disease (5). In other populations in the United States the prevalence is between 1 and 2% of the population. The etiologic factors for type II diabetes mellitus include genetic and environmental factors.

Genetic Factors in Type II DM

Unlike type I DM, there is no strong association between HLA type and type II DM in the general population. There is a strong genetic component in type II DM. About one-third of the offspring of patients with type II DM develop diabetes or abnormalities in glucose tolerance. Concordance of identical twins with type II DM is 90 to 100%, compared to less than 50% in type I DM (4). The mode of inheritance of type II DM is not known but is thought to be multifactorial. The one exception is an uncommon form of type II diabetes, maturity onset diabetes of the young, which occurs in nonobese individuals and is inherited as an autosomal dominant trait (77).

Environmental Factors in Type II DM

Environmental factors play a major role in the development of type II DM. Obesity that induces insulin resistance is a major factor (45). A high percentage of individuals with type II DM are overweight. These patients usually have impaired release of insulin from the islets and the tissue sensitivity to insulin is reduced in various organs.

Pathology of the Islets in Diabetes Mellitus

Type I DM

Early in the disease, there may be an infiltration of lymphocytes and macrophages in the islets, a condition described as insulitis. This lesion has been observed at autopsy in up to two-thirds of diabetics who died within 6 months after the onset of their initial symptoms (26) (Fig. 5.8). Only a small percentage of the surviving islets with beta cells have lymphocytes. Insulitis is usually not seen in longstanding type I diabetes mellitus and it may represent a response to beta cell destruction.

Later, the entire pancreas becomes atrophic. This atrophy, which is commonly observed at autopsy, may be secondary to the lack of high levels of insulin normally found in the vascular connections between the islets and the acinar tissue (38). The islets are markedly decreased in size and number. Beta cells are virtually absent and there is loss of the normal islet architecture.

Type II DM

The islet cell mass is not reduced in type II DM (66). The content of the beta cell mass is normal in size, while the glucagon-producing cell mass is increased. Fibrosis, hyalinization, and amyloid deposits may develop in the islets with chronic type II DM (Fig. 5.9).

Complications of Diabetes Mellitus

The major organs affected as complications of diabetes mellitus include the heart, blood vessels, kidneys, eyes, and nerves. The pathophysiology of these various diseases is not well understood but there are two possible mechanisms: 1) an increase in protein glycosylation such as glycosylated albumin and hemoglobin secondary to unregulated glycosylation because of the high serum glucose concentration, which may change the structure of proteins and alter their function, and 2) abnormalities of the polyol pathway with exposure to high

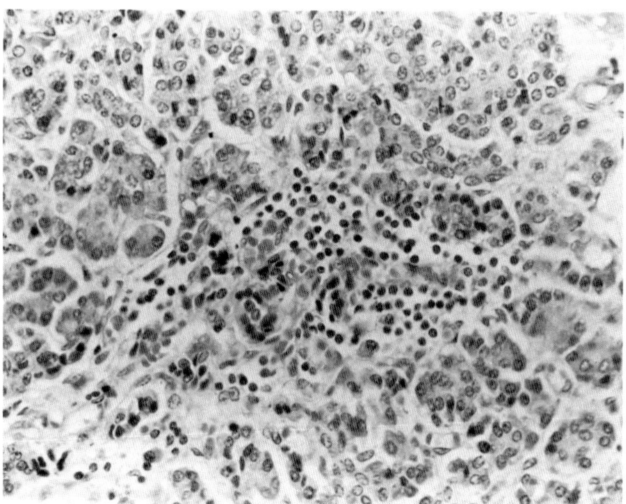

Figure 5.8. Pancreatic islet from a patient with type I diabetes mellitus. Insulitis consisting of an infiltrate of lymphocytes and macrophages into the islet is prominent.

levels of glucose in the retina, kidney, Schwann cells, and aorta. There is increased production of the sugar sorbitol, which is oxidized to fructose. Sorbitols can cause osmotic swelling in the lens and inhibit the Na^+,K^+-ATPase system in nerves. High intracellular concentrations of sorbitol also prevent myoinositol from entering cells. The intracellular deficiency of myoinositol may also contribute to some of the long-term complications of diabetes.

Atherosclerosis

Diabetes increases the extent and severity of atherosclerosis (89). The atherosclerotic risk is greatest in individuals with poorly controlled diabetes mellitus, and this may be related to hypercholesterolemia and hypertriglyceridemia. The coronary, central, and peripheral vessels are commonly involved, which leads to an increased incidence of myocardial infarction and gangrene. Increased platelet adhesiveness is observed in diabetes mellitus and may also be a significant factor in diabetic atherosclerosis.

Nephropathy

Diabetic nephropathy is a major complication of chronic diabetes mellitus. It may account for up to one-fourth of patients receiving long-term renal

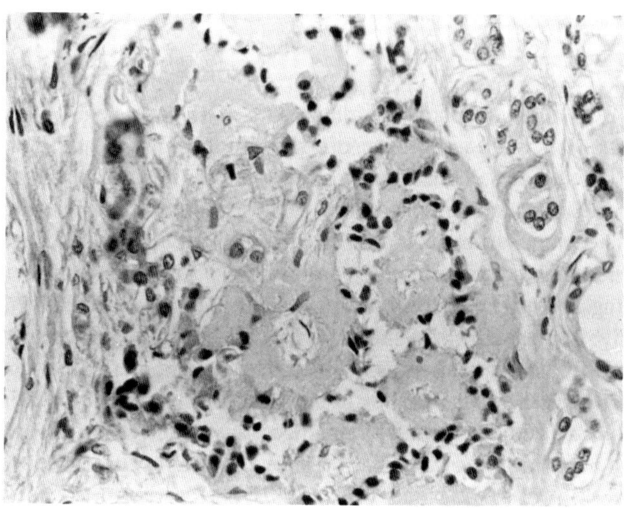

Figure 5.9. Pancreatic islet from a patient with long-standing type II diabetes mellitus. Fibrosis and extensive hyalinization of the islet are present.

CHAPTER 6

Neuroendocrine Cells and Neoplasms of the Gastrointestinal Tract

Anatomy and Development

The endocrine cells of the gastrointestinal tract are of endodermal derivation. The exact cell of origin is unknown but could be a common totipotential stem cell (45). Although the gastrointestinal endocrine cells were among the amine precursor uptake and decarboxylation (APUD) cells that were postulated to have a common neural crest origin (50), the experimental evidence strongly supports an endodermal origin (15,41).

The principle endocrine cell types of the gastrointestinal tract are summarized in Table 6.1. Other less well known cell types with controversial functions have also been described (40). The gastrin-producing G cells are primarily in the gastric antrum (Fig. 6.1) but are also present in the duodenal mucosa and in Brunner's glands of the duodenal bulb. Ultrastructurally, these cells contain secretory granules 190 to 300 nm in diameter. Immunostaining is positive for gastrin-17 and gastrin-34. The granules are present mostly in the basal portions of the G-cell cytoplasm and the apical portion of the cell reaches the glandular lumen. These cells also express chromogranin A protein and mRNA (Fig. 6.2). The somatostatin (δ) cells are found in the stomach and small intestine including Brunner's glands (Fig. 6.3). The secretory granules range from 300 to 400 nm in diameter. Immunostaining shows positive reactivity for somatostatin in these cells. The secretin-producing cells (52) are present in the duodenum and jejunum. These cells have granules 180 to 220 nm in diameter and show positive immunoreactivity for secretin. The I cells, which are also located in the duodenum and jejunum, secrete cholecystokinin (CCK). These cells have secretory granules of 250 to 300 nm and are positive for CCK by immunochemistry. The L cells are present in the mucosa of the small and large intestine. Glucagon-like immunoreactivity has been identified in these cells.

The EC cells are one of the three types of enterochromaffin cells. They are found throughout the small and large intestinal mucosa. They stain positively for serotonin and some also stain for substance P (8). The EC_2 cells are present predominantly in the duodenum. They have large granules 400 nm in

Table 6.1. Endocrine cells of the gastrointestinal tract

Cell type[a]	Location	Peptides	Secretory granules (nm)
G	Antrum	Gastrin	150–300
D	Fundus, antrum, small intestine	Somatostatin	300–400
IG	Duodenum, jejunum	Gastrin	160–220
S	Duodenum, jejunum	Secretion	180–220
I	Duodenum, jejunum	Cholecystokinin	250–300
L	Small and large intestine	Glicentin, entero-glucagon	250–300
EC	Small and large intestine	Serontonin, substance P	200–300
MO	Small intestine	Motilin	200–300

[a]Other cell types of uncertain or unknown functions and their locations include EC_2, jejunum; ECn, antrum, duodenum; K, duodenum, jejunum); D, fundus, antrum, small and large intestine; N, ileum; P, fundus, antrum, small intestine; PG, duodenum; ECL, fundus; and X, fundus.

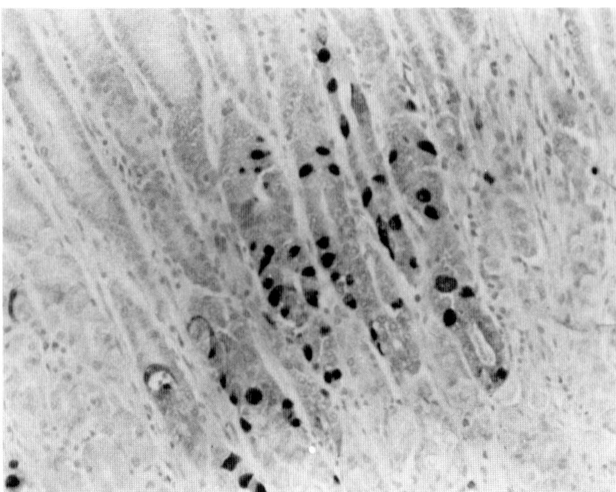

Figure 6.1. Gastrin-producing cells in the antrum of the stomach revealed by immunostaining with an antigastrin antibody.

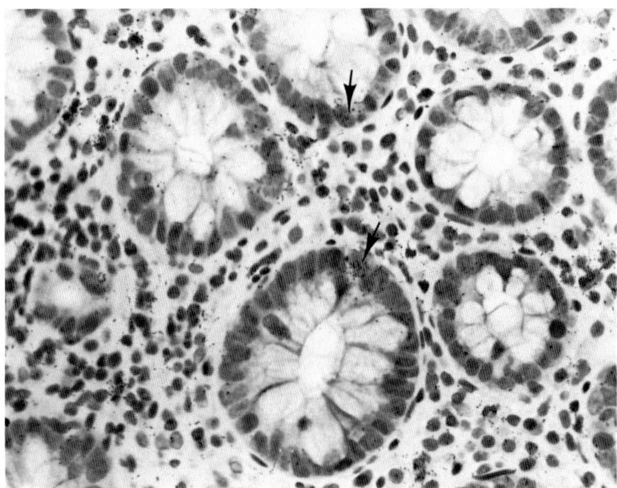

Figure 6.2. Chromogranin A messenger RNA detected in the stomach by in situ hybridization. Many endocrine cells that do not produce defined peptides express chromogranin A protein and mRNA (*arrows*).

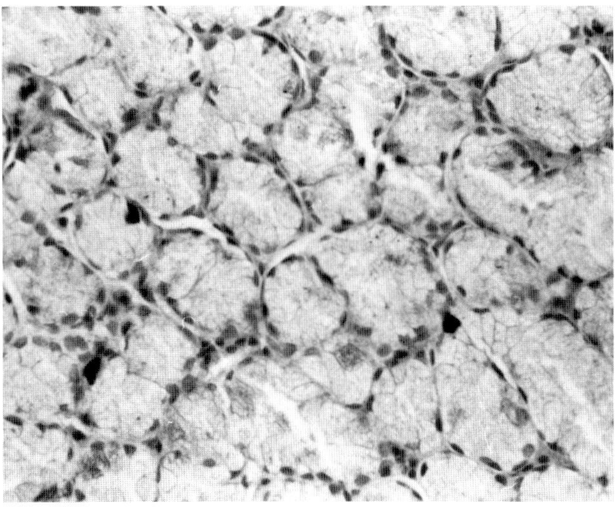

Figure 6.3. Somatostatin-positive cells in Brunner's glands in the duodenum.

Figure 6.4. Chromogranin A-positive cells in the jejunum. Most of these cells also express serotonin.

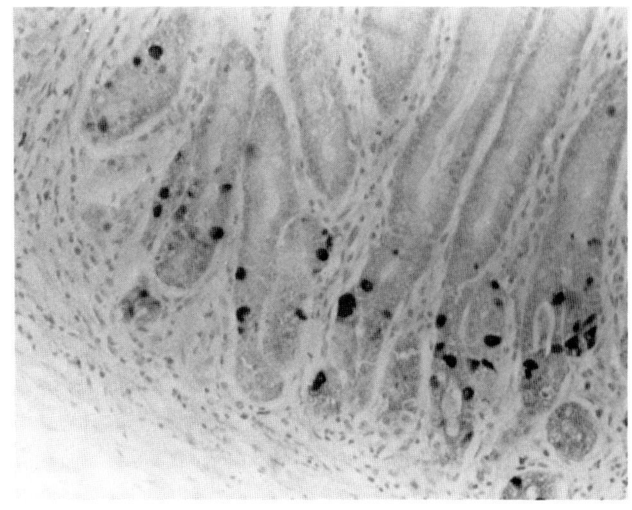

A

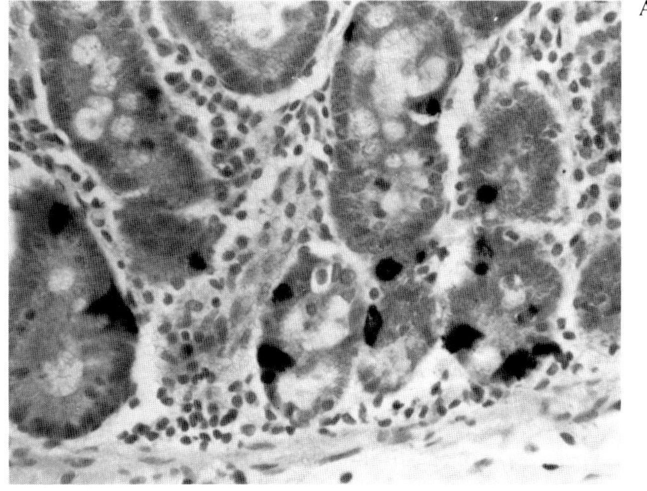

B

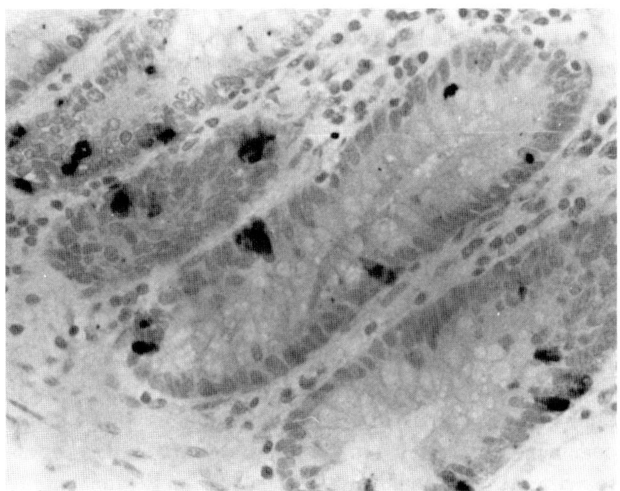

Figure 6.5. (A) Serotonin-positive enterochromaffin cells in the jejunum showing intense immunoreactivity. (B) Serotonin-positive cells in the rectum.

or beyond the liver. Because ovarian carcinoids bypass the portal circulation, the carcinoid syndrome can appear very early with ovarian carcinoids even before metastatic disease occurs.

Neuroendocrine Tumors of the Esophagus

Although there are few endocrine cells in the esophagus compared to other parts of the gastrointestinal tract, endocrine mucosal cells are present in the esophageal cardiac glands of the distal esophagus adjacent to the squamocolumnar junction (54). Ectopic gastric mucosa in the proximal esophagus producing somatostatin, bombesin, gastrin, and vasoactive intestinal polypeptide has also been described (57). Barrett's mucosa, which is a metaplastic change in the esophagus, is associated with neuroendocrine cells including cells staining for serotonin and chromogranin A (7).

Most of the reported neuroendocrine neoplasms of the esophagus have been highly malignant neoplasms (7,54), and carcinoid tumors or well-differentiated neuroendocrine carcinomas are very rare (14).

Grossly, the tumors are large (greater than 4 cm) and may be located anywhere in the esophagus. There is a slightly higher frequency in the mid-esophagus. Most tumors consist of small undifferentiated cells with secretory granules 80 to 200 nm in diameter on electron microscopic examination. The cells look like the cells of small-cell lung carcinomas. These tumors may contain glandular or squamous differentiation. The tumors are argyrophilic and immunochemical stains reveal many peptides present in the metaplastic esophageal mucosa as well as ACTH and calcitonin (54).

The differential diagnosis includes metastatic small-cell carcinoma from the lung and other sites, which can usually be resolved with careful clinicopathological evaluation.

The prognosis for these neuroendocrine tumors is poor, with a median survival of a few months from the time of diagnosis (12). The one unusual case with a long survival reported by Chong and co-workers looked like a trabecular carcinoid (14).

Stomach

Neuroendocrine gastric tumors probably represent 2 to 4% of digestive tract neuroendocrine neoplasms. About 50% of these cases develop metastatic disease (29). Most tumors arise in a background of atrophic gastritis and are usually in the body or fundus of the stomach. Multiple gastric carcinoids may also arise in association with gastric atrophy and achlorhydria. Endocrine cell hyperplasia, dysplastic nodules with cellular atypia, and neoplastic development may be present in a background of gastric atrophy (39). The tumors may be polypoid or nodular.

Microscopically, the tumors have features of foregut carcinoids with a prominent trabecular pattern but may vary from well to poorly differentiated (13,31). Individual cells have eosinophilic cytoplasm and uniform round nuclei. The tumors are usually argyrophilic but may also be focally argentaffin positive. Ultrastructural examination shows secretory granules 200 to 400 nm in diameter. Immunostaining shows immunoreactivity for paraendocrine markers, serotonin, and peptide hormones including somatostatin, motilin, ACTH, pancreatic polypeptide, and gastrin (24, 36) (Figs. 6.7, 6.8). In situ hybridization analyses often show chromogranin A mRNA (Fig. 6.9).

Most G-cell tumors associated with the Zollinger-Ellison syndrome are found in the pancreas (90%), followed by the duodenum (10%). Only a few cases of hypergastrinemia with peptic ulcer are associated with hyperplasia and G-cell tumors.

The 5-year survival for well-differentiated gastric neuroendocrine tumors is around 70%. Polypoid tumors usually have an excellent prognosis.

Small Bowel Duodenum

Most duodenal neuroendocrine tumors are asymptomatic. The duodenal gastrinoma is a notable exception. These tumors may be associated with Zollinger-Ellison syndrome (61a). The tumors are often less than 1 to 5 cm in diameter and are present in the duodenal wall. In one study up to 40% of these tumors were malignant (23). Trabecular and solid growth patterns may be present.

Somatostatin immunoreactivity has been observed in a significant number of duodenal neuroendocrine tumors. Many of these tumors have psammoma bodies and have been termed psammomatous somatostatinomas (17,35). Patients with some of these tumors have von Recklinghausen's disease (17,35). These tumors are often in the

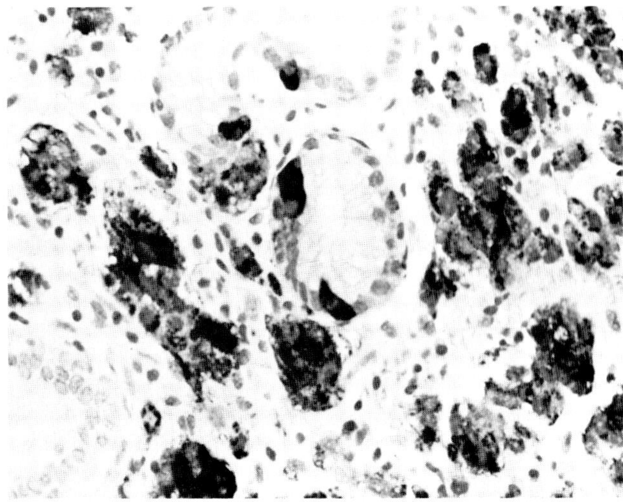

Figure 6.7. Gastric carcinoid tumor immunostained for chromogranin A. The infiltrating tumor cells and nonneoplastic gland cells are positive.

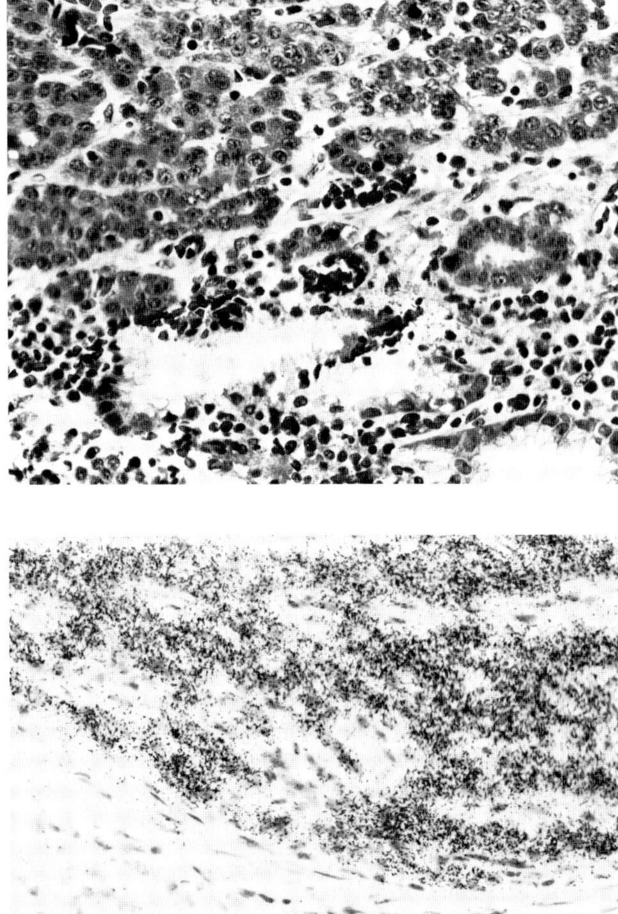

Figure 6.8. Carcinoid tumor of the stomach from a patient with hypergastrinemia. The tumor cells are immunoreactive for gastrin.

Figure 6.9. Gastric carcinoid tumor that was focally positive for gastrin. Most of the tumor cells express the messenger RNA for chromogranin A revealed by in situ hybridization analysis.

A

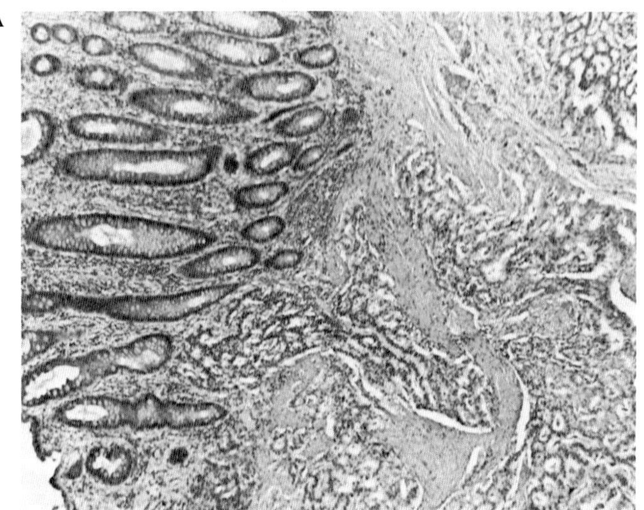

B

Figure 6.18. (A) Rectal carcinoid. The tumor cells are present in the mucosa and infiltrate into the submucosa and muscularis. (B) A prominent trabecular pattern is seen in this rectal carcinoid at a higher magnification.

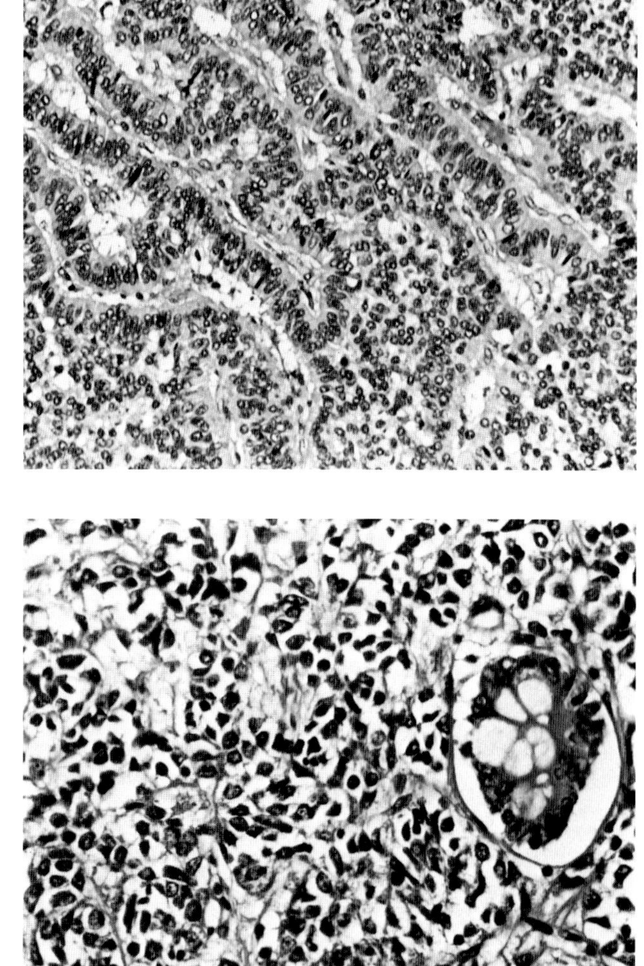

Figure 6.19. Poorly differentiated neuroendocrine carcinoma of the colon. The tumor cells have large nuclei and a small amount of cytoplasm.

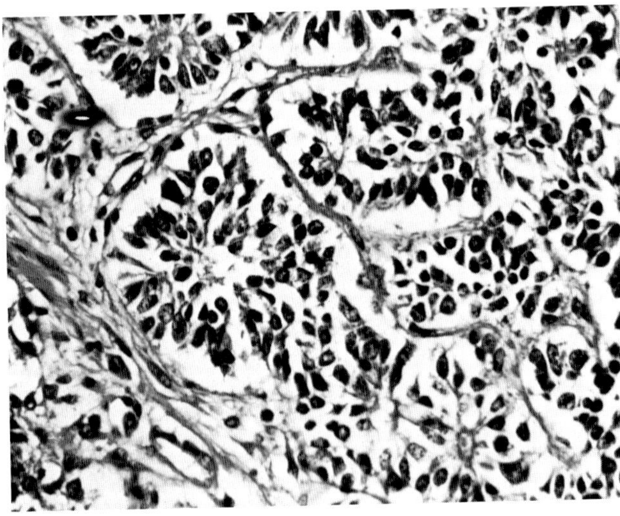

Figure 6.20. This poorly differentiated neuroendocrine carcinoma has nests of tumor cells with large nuclei. Immunoreactive chromogranin A was seen focally, while diffuse staining for neuron-specific enolase was present.

Colon and Rectum

Although carcinoid tumors of the colon constitute a small percentage of gastrointestinal carcinoid neoplasms, the rectum is the third most common site for these neuroendocrine tumors (44). The incidence of metastasis in rectal carcinoids is about 10 to 15% (49). The carcinoid syndrome or clinical evidence of hormonal hypersecretion is rarely associated with rectal carcinoids.

Most tumors are less than 2 cm in diameter and are present in the rectum or lower sigmoid. A trabecular pattern of growth with uniform cells is commonly present (Fig. 6.18A and B). Most tumors are argyrophilic, but the Churukian-Schenk stain (58) is often superior to the Grimelius in showing argryophilic cells in hindgut carcinoids.

Immunochemical staining reveals reactivity for serotonin, somatostatin, pancreatic polypeptide, glicentin, and other peptides. Occasional tumors have been immunoreactive for insulin (6).

Small-cell neuroendocrine carcinomas (poorly differentiated neuroendocrine carcinomas) of the colon and rectum are highly aggressive neoplasms (30) (Figs. 6.19, 6.20). These neoplasms occur in patients between the fourth and eighth decades and metastatic disease is often present at the time of diagnosis. The tumors are generally bulky. Light microscopic features include small to intermediate-size cells arranged in clusters with extensive necrosis. Ultrastructural examination reveals small secretory granules 80 to 250 nm in diameter. Immunoreactivity for neuron-specific enolase, chromogranin A, and Leu 7 is often present (64). In a recent series of 10 cases studied by Wick and co-workers all of the patients died within 11 months of clinical presentation (64).

Tumors with multidirectional differentiation with mixed endocrine and glandular or squamous differentiation have been described in the colon similarly to those in the appendix (30). Immunostaining with broad-spectrum markers often reveals the endocrine component of these neoplasms.

Differential Diagnosis

Carcinoids (well-differentiated neuroendocrine tumors) are almost unique in their appearance and are usually easy to diagnose on H&E sections. With multiple tumors, it may be difficult to decide whether they represent separate primary or metastatic foci. Multifocal primaries are sometimes seen in the ileum but may be present in other areas of the gastrointestinal tract. A careful search for other possible primary tumors at the time of surgery and frozen section evaluation is most helpful.

The atypical carcinoid and poorly differentiated or small-cell neuroendocrine tumors may look like metastatic tumors from the lungs or other sites. Immunochemistry and electron microscopy are not helpful in these cases and clinicopathological correlation is sometimes needed to establish a definitive diagnosis.

Glomus tumors of the stomach or from other

ease at the time of diagnosis of small-cell lung carcinomas (4,24).

Prognosis

The metastatic rate of classical carcinoid tumors is very low. Very few patients die of this disease. Atypical carcinoids are associated with metastatic disease in 60% of cases and 2- to 5-year survivals of 50% have been reported (13).

Small-cell carcinomas that include the intermediate cell type and the oat cell carcinomas are generally associated with a poor prognosis. Aggressive chemotherapy and irradiation therapy have prolonged patient survival in some cases (1).

References

1. Abrams J, Doyle LA, Aisner J (1988) Staging prognostic factors and special considerations in small cell lung cancer. Seimin Oncol 15:261–277
2. Andrew A (1976) An experimental investigation into the possible neural crest origin of pancreatic APUD (islet) cells. J Embryol Exp Morphol 35:577–593
3. Bartter FC, Schwartz WB (1967) The syndrome of inappropriate secretion of antidiuretic hormone. Am J Med 42:790–806
4. Bates M, Hurt R, Levison V, Sutton M (1974) Treatment of oat-cell carcinoma of the bronchus by preoperative radiotherapy and surgery. Lancet 1:1134–1135
5. Bensch KG, Corrin B, Pariente R, Spencer H (1968) Oat-cell carcinoma of the lung; its origin and relationship to bronchial carcinoid. Cancer 22:1163–1172
6. Bensch KG, Gordon GB, Miller LR (1965) Embryology of the diffuse neuroendocrine system and its relationship to the common peptides. Fed Proc 38: 2288–2294
7. Capella C, Hage E, Solcia E, Usellini L (1978) Ultrastructural similarity of endocrine-like cells of the human lung and some related cells of the gut. Cell Tissue Res 186:25
8. Carter D (1983) Small-cell carcinoma of the lung. Am J Surg Pathol 7:787–795
9. Carter D, Eggleston JC (1980) Tumors of the lower respiratory tract. Atlas of Tumor Pathology, Second Series, Fascicle 17. Washington, DC, Armed Forces Institute of Pathology
10. Cureton RJR, Hill IM (1955) Malignant change in bronchiectasis. Thorax 10:131–136
11. Fontaine J, Le Douarin NM (1977) Analysis of endoderm formation in the avian blastoderm by the use of quail-chick chimaeras. J Embryol Exp Morphol 4: 202–222
12. Gould VE (1983) The endocrine lung. Lab Invest 48:507–509
13. Gould VE, Lee I, Warren WH (1988) Immunohistochemical evaluation of neuroendocrine cells and neoplasms of the lung. Pathol Res Pract 183: 200–213
14. Gould VE, Linnoila RI, Memoli VA, Warren WH (1983) Neuroendocrine components of the bronchopulmonary tract: Hyperplasias, dysplasias and neoplasms. Lab Invest 49:519–537
15. Hamid QA, Addis BJ, Spinrgall DR, Ibrahim NBN, Ghatei MA, Bloom SR, Polak JM (1987) Expression of the C-terminal peptide of human pro-bombesin in 361 lung endocrine tumours, a reliable marker and possible prognostic indicator for small cell carcinoma. Virchows Arch [Pathol Anat] 411:185–192
16. Harrison MT, Montgomery DA, Ramsey AS, Robertson JH, Welborn RB (1953) Cushing's syndrome with carcinoma of bronchus and with features suggesting carcinoid tumor. Lancet 2:23–25
17. Lauweryns JM, Cokelaere M (1973) Intrapulmonary neuro-epithelial bodies: Hypoxia-sensitive neuro(chemo-)receptors. Experientia 29:1384–1386
18. Lauweryns JM, Cokelaere M, Theurnynck P (1973) Serotonin producing neuroepithelial bodies of rabbit respiratory mucosa. Science 180:410–413
19. Lauweryns JM, Peuskens JC (1972) Neuro-epithelial bodies (neuro-receptor or secretory organs) in human infant bronchial and bronchiolar epithelium. Anat Rec 172:471–482
20. Lauweryns JM, Peuskens JC, Cokelaere M (1970) Argyrophilic, fluorescent and granulated (peptide and amine producing?) AGF cells in human infant bronchial epithelium; light and electron microscopic studies. Life Sci 9:1417–1429
21. Lauweryns JM, Van Ranst L, Lloyd RV, O'Connor DT (1987) Chromogranin in bronchopulmonary neuroendocrine cells. Immunocytochemical detection in human, monkey, and pig respiratory mucosa. J Histochem Cytochem 35:113–118
22. LeDourin N (1982) The Neural Crest. Cambridge, England, Cambridge University Press
23. Lee I, Gould VE, Moll R, Wiedenmann B, Franke WW (1987) Synaptophysin expression in pulmonary neuroendocrine cells, neuroepithelial bodies and neuroendocrine neoplasms. Differentiation 34:115–125
24. Levison V (1980) Pre-operative radiotherapy and surgery in the treatment of oat-cell carcinoma of the bronchus. Clin Radiol 31:345–348

25. Linnoila RI, Mulshine JL, Steinberg SM, Funa K, Matthews MJ, Cotelingam JD, Gazdar AF (1988) Neuroendocrine differentiation in endocrine and neuroendocrine lung carcinomas. Am J Clin Pathol 90:641–652
26. Linnoila RI, Nettesheim P, DiAugustine RP (1981) Lung endocrine-like cells in hamsters treated with dimethylnitrosamine: alterations *in vivo* and in cell culture. Proc Natl Acad Sci USA 78:5170–5174
27. Memoli VA, Linnoila I, Warren WH, Rios-DaLenz J, Gould VE (1983) Hyperplasia of pulmonary neuroendocrine cells and neuroepithelial bodies. Lab Invest 48:57A (abstract)
28. Palisano JR, Kleinerman J (1980) APUD cells and neuroepithelial bodies in hamster lung; methods, quantitation and response to injury. Thorax 35:363–370
29. Pearse AGE (1969) The cytochemistry and ultrastructure of polypeptide hormone producing cells of the APUD series and the embryologic, physiologic and pathologic implications of the concept. J Histochem Cytochem 17:303–313
30. Pearse AGE, Takor-Takor T (1979) Embryology of the diffuse neuroendocrine system and its relationship to the common peptides. Fed Proc 38:2288–2294
31. Reznik-Schuller H (1976) Ultrastructural alterations of nonciliated cells after nitrosamine treatment and their significance for pulmonary carcinogenesis. Am J Pathol 85:549–554
32. Saccomanno G, Archer VE, Auerbach O, Kuschner M, Saunders RP, Klein MG (1971) Histologic types of lung cancer among uranium miners. Cancer 27: 515–523
33. Stahlman MT, Kasselberg AG, Orth DN, Gray ME (1985) Ontogeny of neuroendocrine cells in human fetal lung. II. An immunohistochemical study. Lab Invest 52:52–60
34. Tischler AS (1978) Small cell carcinoma of the lung: cellular origin and relationship to other neoplasms. Semin Oncol 5:244–250
35. Tsutsumi Y, Osammura RY, Watanabe K, Yanaihara N (1983) Immunohistochemical studies on gastrin-releasing peptide and adrenocorticotropic hormone-containing cells in the human lung. Lab Invest 48: 623–632
36. Yesner R (1983) Small cell tumors of the lung. Am J Surg Pathol 7:775–785

Table 8.1. Etiology of chronic adrenocortical insufficiency

I. Primary
Common
Idiopathic (autoimmune) atrophy
Granulomatous infections
Uncommon
Congenital
Metastatic carcinoma
Amyloidosis
Hemochromatosis
Adrenoleukodystrophy
Bilateral adrenalectomy
Wolman's disease
II. Secondary
Pituitary disorders
Suppression of pituitary by exogenous glucocorticoids
III. Tertiary
Hypothalamic disorders

Adrenocortical Insufficiency

Adrenocortical insufficiency may develop acutely as a life-threatening condition commonly associated with bacterial infections. If it is not diagnosed correctly and treated, the patient usually develops vomiting, hypotension, diarrhea with subsequent shock, and death. Chronic adrenocortical insufficiency has many causes, ranging from an immunologic etiology to surgical adrenalectomy (Table 8.1). Chronic insufficiency is commonly associated with fatigue, muscle weakness, anorexia, weight loss, hypotension, and hyperpigmentation of the skin and mucous membrane secondary to excessive production of proopiomelanocortin-related peptides.

Waterhouse-Friderichsen Syndrome

This syndrome, which occurs most often in children, is characterized by an acute onset of sepsis associated with severe bacterial infections leading to acute hemorrhagic adrenal infarction (32). The organisms commonly associated with this condition include meningococcus and pneumococcus. *Hemophilus influenzae* and *Staphylococcus* may also be the etiologic organisms. The disease is probably due to intravascular clotting such as occurs in septic shock due to bacteremia and endotoxemia. Patients develop high fever, purpuric rash, and hypotension, which can result in vascular collapse and irreversible shock.

The adrenal glands are diffusely or focally discolored, dark red, and slightly enlarged. There is extensive hemorrhagic necrosis of the cortex, especially the zonae reticularis and fasciculata, and the medulla may also be involved (Fig. 8.5). Neutrophilic infiltrate may be seen in the glands if enough time has transpired between the acute infarcts and examination of the adrenals.

Acute adrenocortical insufficiency may also be seen in patients with chronic insufficiency combined with acute stresses such as trauma or sepsis. Rapid withdrawal of steroids from patients with chronic steroid treatment may also lead to acute adrenocortical insufficiency.

Chronic Adrenocortical Insufficiency

Idiopathic or primary adrenal cortical atrophy is the most common cause of Addison's disease today. It accounts for about 70% of the cases of Addison's

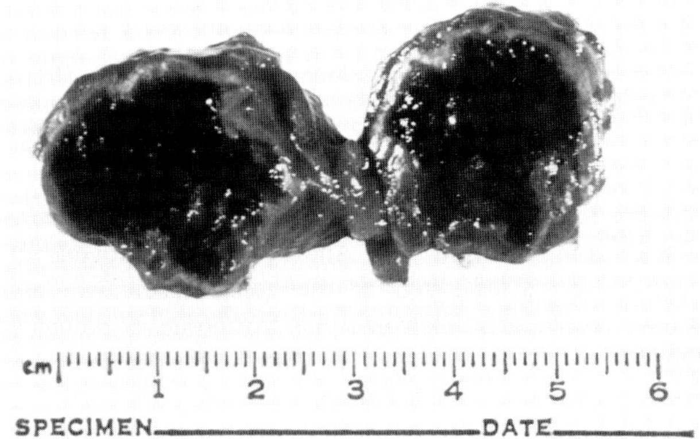

Figure 8.5. Adrenal cortical insufficiency secondary to acute hemorrhagic adrenal infarction. A rim of uninvolved adrenal cortex is present.

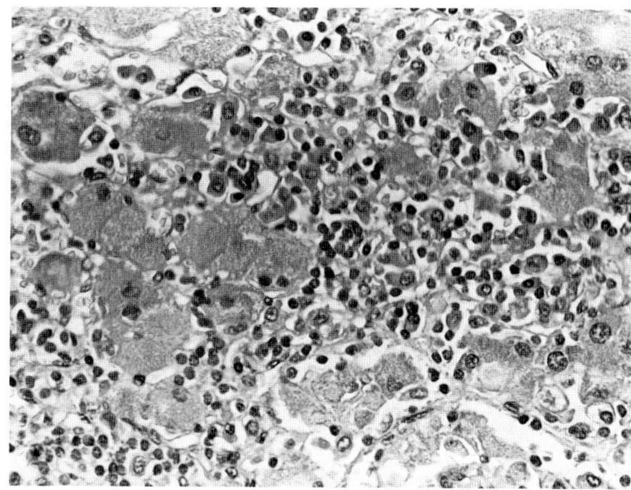

Figure 8.6. Primary adrenal cortical atrophy (Addison's) showing a marked lymphocytic infiltrate and some residual adrenal cortical cells.

disease in the United States; in contrast, early in the 20th century tuberculosis was the principal cause of Addison's disease (37,65). The incidence of primary adrenal atrophy is higher in women with a ratio of 25 : 1 (women : men). It is most common in younger patients.

The presence of circulating adrenal autoantibodies directed against adrenal cortical microsomes and mitochondria in over 50% of patients with this disease is suggestive evidence for an autoimmune etiology (6,65). Autoantibodies against thyroid, gastric mucosa, and intrinsic factor are often also present. Some patients may have Schmidt's syndrome, which is the association of Addison's disease with primary myxedema, or insulin-dependent diabetes mellitus.

More than 90% of the adrenal cortex must be destroyed before the typical signs of insufficiency become apparent. The glands in Addison's disease are shrunken and irregular, weighing 1 to 3 g. The histology of the glands reveals an intact medulla with extensive cortical fibrous connective tissue. A few residual nests of cortical cells may be present. Lymphocytic infiltrates are commonly present (Fig. 8.6). If residual cortical cells are present, they are commonly lipid depleted.

Granulomatous Infection

The most common causes of granulomatous infection of the adrenal glands in the United States are tuberculosis and histoplasmosis. Tuberculosis was the most common cause of Addison's disease in the United States until a few decades ago.

The glands are enlarged and irregular with adherence to adjacent structures such as the kidneys. Histologically, few adrenal cortical cells can be recognized. Residual cells are usually subcapsular when present. Fibrosis, caseous necrosis, granulomas, and lymphocytes are common. Dystrophic calcification is sometimes seen, depending on the duration of the disease.

Rare Causes of Adrenocortical Insufficiency

Various conditions including adrenal hyperplasia and adrenogenital syndrome, which will be discussed later, may lead to adrenocortical insufficiency in neonates and infants. Anencephalic neonates have hypoplastic adrenal glands. Cytomegalic adrenal hyperplasia may be a focal or diffuse condition in which cells of the fetal cortex have bizarre nuclear features and abundant eosinophilic cytoplasm and are lipid depleted. The cells may be up to 100 μm in diameter. This condition may be associated with Beckwith's syndrome (9). The adrenal glands are usually less than 1 g in weight. Diffuse involvement can be associated with adrenocortical insufficiency.

Amyloidosis secondary to chronic granulomatous infection (secondary amyloidosis) may lead to enlargement of the glands due to interstitial deposits of amyloid, which produce a waxy gross

A

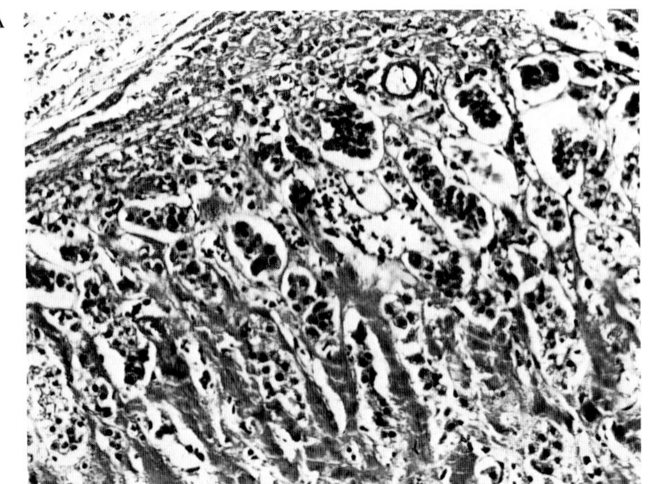

B

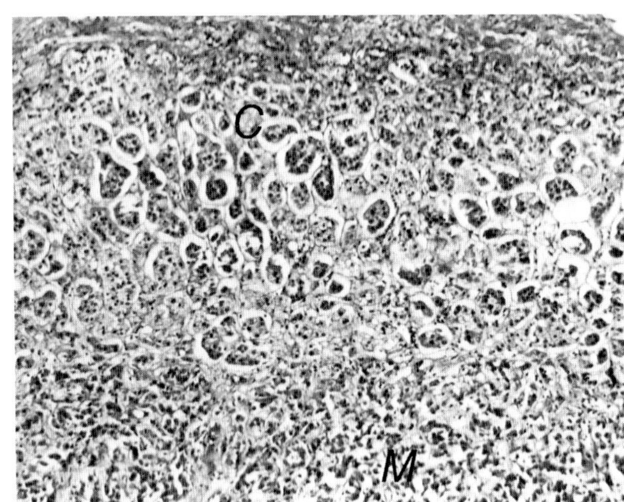

Figure 8.7. (A) Amyloidosis of the adrenal. The etiology is unknown. Extensive amyloid has been deposited in the cortex and there is marked adrenal cortical atrophy. (B) Adrenal cortical atrophy secondary to chronic high-dose glucocorticoid therapy. The cortical zones (*C*) are reduced in size while the medulla (*M*) is relatively unaffected.

appearance (Fig. 8.7). Addison's disease may develop when there is extensive amyloid deposit in the inner cortical zones (40). Focal amyloid deposits secondary to plasma cell disorders usually do not lead to Addison's disease.

Hemochromatosis and hemosiderosis can lead to extensive hemosiderin deposits, especially in the zona glomerulosa. Addison's disease is rarely associated with these conditions. Metastatic neoplasms of the adrenal gland may occasionally produce Addison's disease (101).

Adrenoleukodystrophy is associated with adrenocortical atrophy in addition to demyelination in the brain and mental retardation (89). This is a rare sex-linked disorder that occurs in males especially during the first decade of life. It is a generalized lipid storage disorder in which there is a defect in steroid hormone metabolism with increased production of very long chain fatty acid (95). The adrenal cortex is markedly atrophic and the residual inner cortical cells, when present, are enlarged and have pale cytoplasm. Ultrastructural studies reveal filaments and fibrillar material in the cytoplasm of these cells (95). Antibodies to adrenal tissues are not present. The adrenal medulla is not affected; however, other steroid-producing cells such as the Leydig cells also have fibrillary cytoplasmic deposits (95). In spite of treatment of Addison's disease, the brain disease progresses inevitably to death.

Wolman's disease is a rare familial disorder

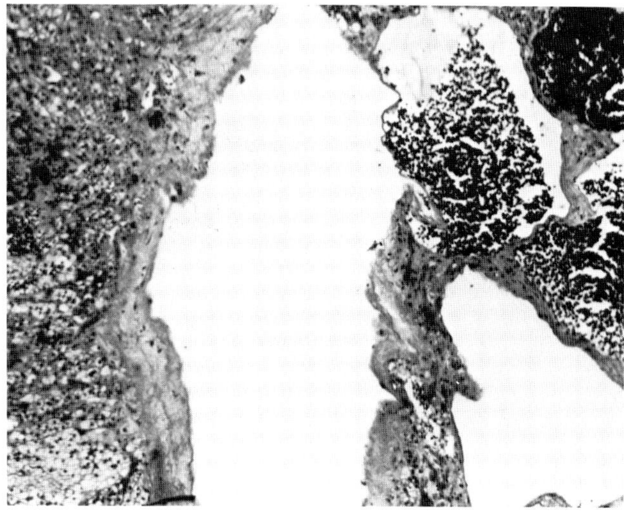

Figure 8.8. Adrenal cortical cysts. The patient presented with an adrenal mass and preoperative studies showed an adrenal cyst. Histologic sections revealed dilated spaces, some lined by endothelial cells, and recent and old hemorrhage.

caused by deficiency of lysosomal enzyme activity. The adrenals are enlarged and calcified. Hepatomegaly with foam cells filled with cholesterol and triglycerides is present. Patients usually die within a few months (124).

Secondary and Tertiary Adrenocortical Insufficiency

Adrenocorticotropin (ACTH) and corticotropin releasing hormones (CRH) produced by the pituitary and hypothalamus, respectively, are needed to maintain the functional integrity of the adrenal cortex. Hypothalamic and/or pituitary failure leads to atrophy of the adrenal cortex with a normal residual medulla. Hypopituitarism is commonly associated with atrophy of the thyroid gland and gonad as well. Primary adrenocortical insufficiency is associated with increased skin pigmentation while secondary insufficiency is not, since the peptides from the proopiomelanocortin complex including melanocyte-stimulating hormone are not being produced.

Glucocorticoid-Induced Atrophy

High levels of exogenous or endogenous corticosteroids can lead to suppression of the hypothalamo-pituitary axis and produce adrenocortical atrophy (Fig. 8.7B). This form of adrenocortical insufficiency can also result from unilateral removal of a glucocorticoid-producing neoplasm that had previously produced atrophy of the contralateral adrenal gland.

Adrenal cortical atrophy with reduction in the size and lipid content of the cortical cells is present. The severity of the condition depends on the dosage and duration of the treatment; adrenal insufficiency lasting for several days develops after treatment with over 100 mg of cortisol daily for more than 1 week (54).

Nonneoplastic Enlargement of the Adrenals

Cysts

Most adrenal cysts are small and are found incidentally at surgery or at necropsy. Occasionally large cysts up to a few centimeters in size may become symptomatic and require surgery (Fig. 8.8).

Grossly, the cysts contain brown or bloody fluid. The wall is often less than 1 cm thick and consists of fibrous tissue with areas of hemosiderin deposits after calcification. Yellow islands of cortical tissues may be seen. Lymphangiectatic cysts are multiloculated and contain a clear or slightly milky fluid.

Microscopically, fibrous tissue, hemosiderin deposition, macrophages, and focal calcification are often present. Clusters of adrenal cortical cells may be present in the wall. Lymphangiectatic cysts have a flattened endothelial lining with fibrous tissue and smooth muscle bundles in the wall (49). Pseudocysts do not have a recognizable lining layer (70a).

Microcysts are often seen in the cortex of infarcts and may be related to stress (84). These cysts are lined by adrenocortical cells and contain red blood cells, protein, and necrotic cells in the lumen (84).

Hemorrhage and Necrosis

Small foci of hemorrhage and necrosis of the adrenal may be present secondary to hypotension and hypoperfusion. More extensive hemorrhage may be associated with the Waterhouse-Friderichsen syndrome, trauma, disseminated intravascular coagulation, and septicemia.

Nonneoplastic Adrenocortical Nodules

Nonneoplastic nodules are a common finding in the adrenal cortex, especially at autopsy. There is an increased incidence of adrenal cortical nodularity with advancing age (80). The nodules are often bilateral. These nodules are more common in individuals with hypertension and vascular disease. Dobbie has proposed that the prominent hyaline vessels seen in these adrenals may be related to ischemic atrophy and subsequent compensatory nodular hyperplasia (22). However, these nodules are not associated with increased adrenocortical function in cultured adrenal cortical cells (86).

The nodules are usually less than 3 cm in size. Multiple nodules are always present. They are well demarcated but not encapsulated and appear continuous with the adjacent adrenal gland. Nodules may be yellow to brown.

Microscopically, the cells are arranged in cords and nests separated by fibrovascular trabeculae. The cortical cells around the central veins are often involved in nodular change. The cells of nodules often contain abundant lipid even when the adjacent cells are lipid depleted. Occasionally, fibrosis and hemorrhage may be present in the nodules.

The differential diagnosis includes adenomas and multinodular hyperplasia. Since incidental nodules are nonfunctional, there is usually no evidence of inactivity or hyperactivity in the adjacent cortex. Unlike adenomas, these nodules are multiple, do not have a distinct capsule, and are continuous with adjacent cortical cells. Clinically, there is no evidence of excessive glucorcorticoid production in patients with incidental nodules (18,34).

Table 8.2. Etiology of hyperfunctional adrenal cortex

Hyperplasia
Congenital adrenal hyperplasia
Cushing's disease
Ectopic ACTH/CRH production
Hyerplasia with hyperaldosteronism
Multinodular hyperplasia with hypercortisolism
Bilateral pigmented nodular adrenocortical disease
Neoplasms
Adenomas
Aldosterone producing
Glucocorticoid producing
Virilizing
Black adenoma
Nonfunctional
Carcinomas
Functional and nonfunctional

Hyperfunctioning of the Adrenal Cortex

Hyperfunctional adrenal cortex can have many etiologies (Table 8.2). It may lead to distinct clinical manifestations such as Cushing's syndrome, Conn's syndrome (primary aldosteronism), virilization or feminization conditions, or a combination of these. Hyperplasia and primary benign and malignant neoplasms are the most common causes of these disorders, though ectopic ACTH production and idiopathic causes may also lead to hyperfunction of the adrenal cortex.

Congenital Adrenal Hyperplasia

These disorders are caused by specific enzymatic defects in the steroid hormone biosynthetic pathway. Most of the patients are female and these conditions are most common in the pediatric age group. The most common defect is absence of the 21-hydroxylase enzyme, which accounts for more than 90% of cases. This deficiency leads to impaired cortisol production with increased ACTH output because of a decrease in the negative feedback to the pituitary. It also causes an increase in androgen biosynthesis, leading to virilism and salt-losing features. The second most common defect is 11β-hydroxylase deficiency, which accounts for about 5% of cases. A deficiency of this enzyme leads to impaired production of cortisol and corticosterone, resulting in hypertension because of mineralocortical accumulation (11-deoxycorticosterone and 11-deoxycortisol). Virilism can also develop in these individuals because of increased androgen production. Other rare enzyme deficiencies include 17-hydroxylase deficiency leading to immature female phenotype and hypertension, 3 beta oℓ-dehydrogenase deficiency leading to sexual ambiguity and salt-losing features, and 20,22 desmolase complex deficiency leading to complete adrenal insufficiency and failure of development and differentiation of male gonads and genitalia.

The adrenals are enlarged, diffusely brown, and have a cerebriform appearance with most enzyme defects associated with congenital adrenal hyperplasia. The glands may weigh up to 30 g each (76, 110). With the 20,22 desmolase complex defect, the adrenals are nodular and yellow due to extensive cholesterol accumulation.

Microscopically, the glands have extensive infoldings with thickened cortex. Compact cells are dominant. The zona glomerulosa may also be hyperplastic when there is salt-losing syndrome secondary to decreased or absent aldosterone production from corticosterone.

These conditions are usually detected at birth or shortly after birth when the defects are severe. With mild defects, the diagnosis may not be made until around puberty. Treatment with cortisol is often effective.

Adults with 21-hydroxylase deficiency may have testicular enlargement with prominent Leydig cells (83).

Hyperplasia Secondary to Pituitary Disorders

The most common cause of adrenal cortical hyperplasia (80%) is pituitary hyperfunctioning with excessive production of ACTH. ACTH-producing adenomas of the pituitary are most frequent, but rare cases may be due to ACTH cell hyperplasia (70). Most patients with pituitary ACTH adenomas are in the third and fourth decades and there is a female-to-male ratio of 3 to 1. In children the sex ratio is about equal. Most children with pituitary adenomas leading to Cushing's disease are older than 10 years.

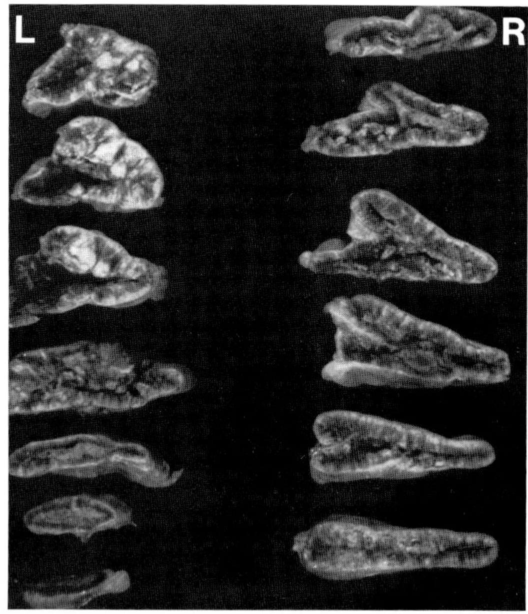

Figure 8.9. Secondary adrenal hyperplasia in a patient with Cushing's disease and a pituitary adenoma. The left adrenal weighed 5 g and the right adrenal 7.9 g. The cortex is brown-yellow and a few cortical nodules are present.

The adrenal glands each weigh 5 to 12 g (Fig. 8.9). Diffuse enlargement is common, but larger glands may be slightly nodular. The inner cortex is prominent and brown while the outer cortex is yellow. When nodules are present they are irregularly distributed and are yellow to brown.

Microscopically, the zone reticularis is widened and the compact cells are often sharply demarcated with an outer zone of clear cells (Fig. 8.10). A prominent zona reticularis may be seen with Cushing's disease even with glands of normal weight and may be the only histologic clue to correlate with marked biochemical and clinical findings.

Ultrastructural studies often show decreased lipid globule size and increased smooth endoplasmic reticulum (91).

Ectopic ACTH/CRH Production

Ectopic ACTH production, most often by lung tumors, can lead to diffuse enlargement of the adrenal glands. Serum ACTH levels are usually

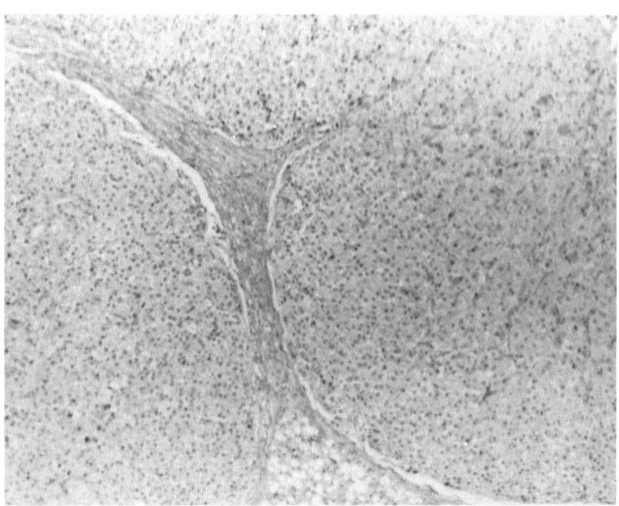

Figure 8.10. Adrenal cortical hyperplasia secondary to Cushing's disease. A nodular pattern is present.

higher than with pituitary disease. CRH production by a tumor can lead to stimulation of pituitary ACTH cell hyperplasia with a histological picture in the adrenals similar to that seen with Cushing's disease. In most cases of excessive CRH production, the tumor is also producing ACTH (97).

Grossly, the adrenal glands are symmetrically enlarged. The cortex is diffusely widened and the glands may weigh between 12 and 30 g, which is larger than with pituitary Cushing's disease and may be related to the higher serum ACTH levels. The cortex is regularly thickened and brown throughout and usually measures at least 4 mm.

Histological features include straight columns of hypertrophical compact cells without nodule formation or prominent lipid-filled cells. The zona glomerulosa may be normal or compressed by the compact cells. Metastatic deposits of the ACTH-producing neoplasm may be present in the cortex and the cells adjacent to the metastatic tumor may show marked nuclear pleomorphism from a paracrine effect of ACTH.

Hyperplasia with Conn's Syndrome

Occasional patients with Conn's syndrome or hyperaldosteronism may have hyperplasia of the adrenals without any distinct adenomas. Removal of a single adrenal gland may not improve the clinical signs and symptoms (28,80).

Gross examination of the adrenals may not reveal any abnormalities. Rarely a slight prominence of the cortex may be present.

Microscopically, the zona glomerulosa cells may be present in small groups in a triangular formation toward the central vein with the narrow tip toward the vein. Zona fasciculata-type cells may be prominent (Fig. 8.11). Increased interstitial fibrous tissue and spironolactone bodies in the cytoplasm of the glomerulosa cells may be present in patients treated with spironolactone (Fig. 8.12). Extensive sampling of the adrenal glands should be done to look for a small adenoma.

Multinodular Adrenal with Hypercortisolism

This is a rare condition seen more often in adults. There is bilateral hyperplasia of the adrenals with multiple large nodules along with elevated ACTH levels (15,41,87).

Grossly, the nodules range from 1–2 mm to 3 cm or more in diameter and involve most of the adrenal cortex. The glands usually weigh between 30 and 50 g each. The nodules are yellow-brown.

Microscopically, compact and clear cells are admixed in the various nodules. The cortex between the nodules is also hyperplastic and the cells are histologically similar to those seen in the large nodules. Focal areas with pleomorphic cells containing bizarre nuclei may be present within or between the nodules. Most of the nodules are com-

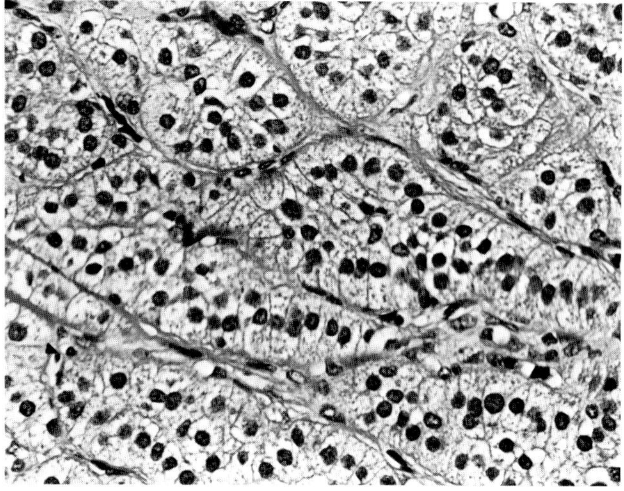

Figure 8.11. Adrenal cortical hyperplasia in a patient with Conn's syndrome or hyperaldosteronism. A distinct adenoma was not present in the gland. Histologic sections revealed zona fasciculata-type cells that are rich in lipid.

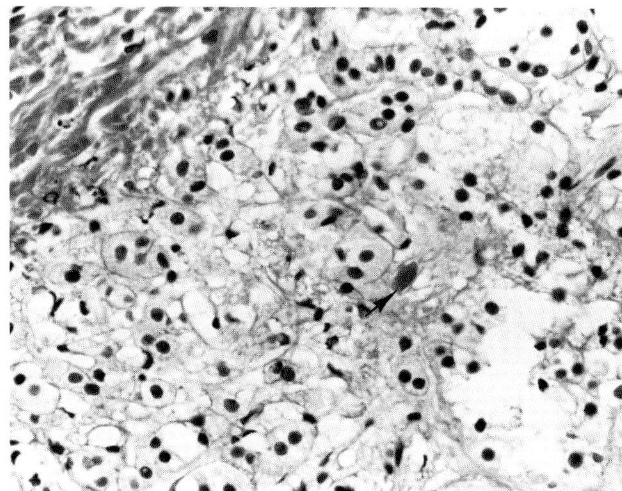

Figure 8.12. Spironolactone body (*arrow*) in the cytoplasm of adrenal cortical cells from a patient with hyperaldosteronism who was treated with spironolactone. These inclusions are seen occasionally in untreated patients.

posed of mixed compact, clear, and intermediate-type cells, unlike the occasional nodules of Cushing's disease, which are composed of clear cells.

Bilateral adrenalectomy is usually curative in these patients as with Cushing's disease.

Bilateral Pigmented Nodular Adrenocortical Disease

This disorder is associated with Cushing's syndrome, although the serum ACTH levels may be normal or slightly below normal. This disease is also known as nodular dysplasia, polymicroadenomatosis, and micronodular adrenal disease (55,77). Several workers have reported the association of this adrenal lesion with a familial syndrome (12,99,102). This rare disorder occurs in children and young adults, with more than 40 reported cases (12). The adrenal hyperactivity is probably autonomous, since dexamethasone does not suppress the hypercortisolism. Patients may have cardiac myxomas, skin pigmentation, skin myxomas, myxoid inflammatory fibroadenomas, testicular Serotoli cell tumors, and occasionally pituitary adenomas. Cardiac myxomas are the most common association and were present in 72% of 40 cases reviewed (12). The etiology of this condition is not known.

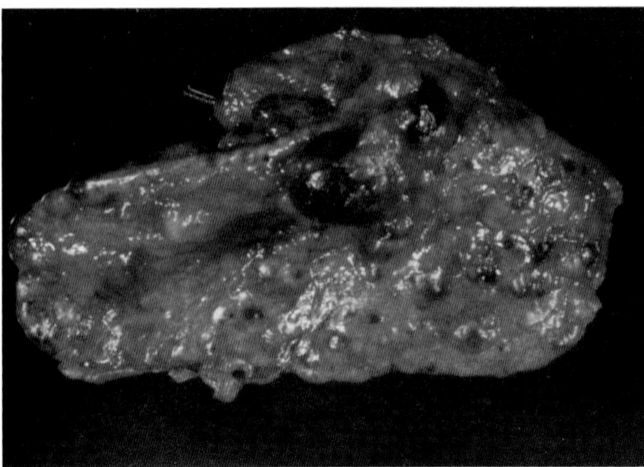

Figure 8.13. Gross photograph of an adrenal gland with bilateral pigmented adrenocortical disease. Both adrenals were of normal weight. Numerous black spherical nodules less than 4 mm in diameter are present on the surface.

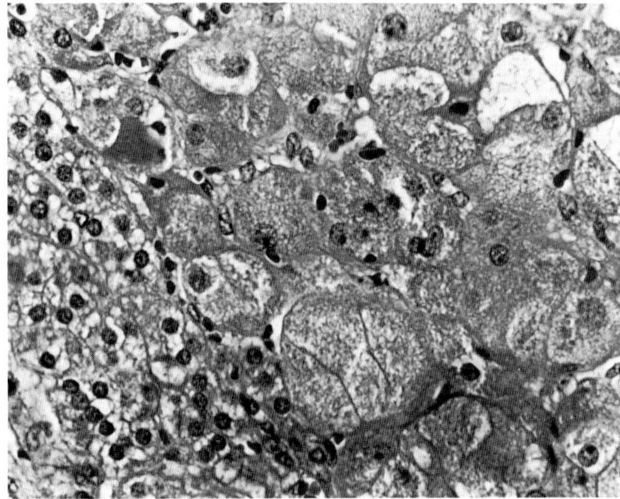

Figure 8.14. Histologic examination of the adrenal in Fig. 8.13 revealed large cells with eosinophilic granular cytoplasm. Nuclear pleomorphism and lipofuscin-type pigment are present in these cells.

Grossly, the adrenal glands may be normal or slightly increased in weight up to 9 g (Fig. 8.13). The glands are small and are dark red-black or yellow due to studding of the surface by small spherical nodules less than 4 mm in diameter. The nodules are most prominent at the corticomedullary junction. Atrophic yellow cortex may be present between the nodules.

Microscopically, the nodules contain cells with eosinophilic granular cytoplasm and large cells with granular or vacuolated cytoplasm. Some cells have granular brown pigment (Fig. 8.14). The pigment is PAS, Fontana-Masson, and Ziehl Neelsen positive and Prussian blue negative consistent with lipofuscin. Focal fatty or myeloid metaplasia is often seen in some nodules. The cortex between the nodules is markedly atrophic. The adrenal medulla is usually not involved except for compression by the adjacent nodules.

Ultrastructural features include prominent lysosomes and pigment bodies in cells within the nodules with tubulovesicular cristae mitochondria and abundant smooth endoplasmic reticulum.

Adrenal Cortical Neoplasms Associated with Specific Syndromes

Hyperaldosteronism

Most tumors causing hyperaldosteronism or Conn's syndrome are benign (17,33,82). A combination of hyperplastic nodules and a dominant nodule is com-

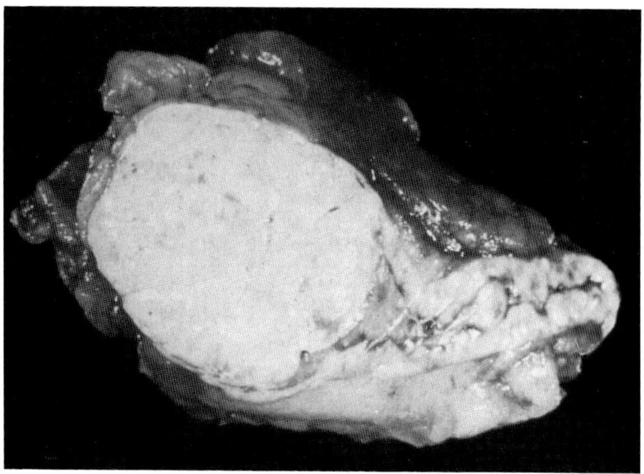

Figure 8.15. Gross specimen from a patient with primary hyperaldosteronism due to an adenoma. The tumor was yellow and about 3 cm in diameter.

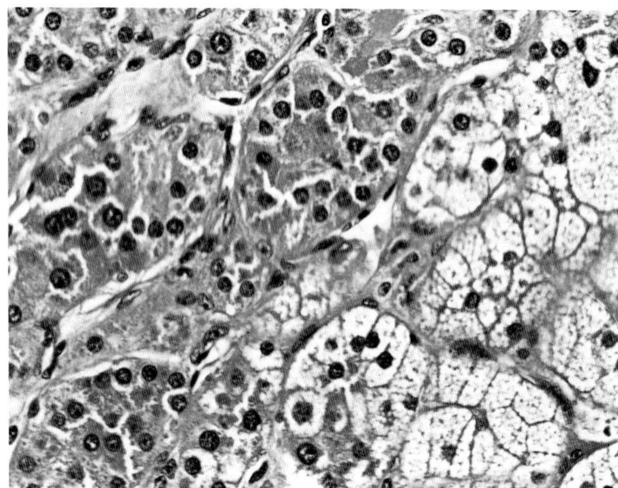

Figure 8.16. Histologic specimens of an adrenal adenoma for a patient with primary hyperaldosteronism. Zona glomerulosa cells, zona fasciculata cells, with "hybrid" cells showing features of both zones are present.

mon. The disease is characterized by hypertension, potassium wasting with neuromuscular dysfunction including muscle cramps, weakness, and tetany. The plasma renin is low and aldosterone levels are elevated. In primary hyperaldosteronism the elevated levels of aldosterone are caused by hyperplasia and adenomas but only rarely by carcinoma. In secondary hyperaldosteronism the increased aldosterone levels and hypertension are related to increased renin synthesis by the kidneys secondary to renal ischemia or a renin-secreting neoplasm from the justaglomerular apparatus of the kidney. Aldosteronomas are more common in women than men (ratio of 3 to 1) and usually develop during the fourth and fifth decades of life.

Adenomas

Most adenomas are less than 3 cm in diameter and weigh less than 6 g. The average size is about 1.5 cm in diameter. These lesions may occur alone or in combination with hyperplastic nodules. The tumors are yellow and the nonneoplastic adrenal is often normal or there may be hyperplasia of the zona glomerulosa (Fig. 8.15).

Microscopically, the cells are of zona fasciculata and glomerulosa type. A hybrid cell described by Symington (111) is often seen (Fig. 8.16). Hybrid cells have features of zona glomerulosa and fasciculata-type cells. Spironolactone bodies, which are eosinophilic multilaminated bodies, are often seen in the cytoplasm of tumors from patients treated with spironolactone before surgery (Fig.

8.12). However, similar bodies have also been seen in patients with cortical adenomas not treated with spironolactone.

Carcinomas

Adrenocortical carcinomas causing primary hyperaldosteronism are rare (31,46). Most of the reported cases have been large tumors weighing up to 2000 g. Areas of hemorrhage, necrosis, and mitoses are commonly seen.

Cushing's Syndrome

Increased production of cortisol from an adrenal cortical adenoma or carcinoma can lead to Cushing's syndrome. Patients usually have a centripetal distribution of adipose tissue leading to truncal obesity, moon faces, and buffalo hump. In addition, easy bruisability, muscle weakness, skin atrophy, and insulin-resistant diabetes mellitus are often present. Patients with adrenal cortical neoplasms associated with Cushing's syndrome may also have virilism or feminization.

Hyperplasia of the adrenals is the most common cause of Cushing's syndrome. About 20% of adult patients with Cushing's syndrome have a neoplasm of the adrenal cortex. In children, neoplasms account for 50 to 60% of the cases of Cushing's syndrome. The female-to-male ratio in adults and children with Cushing's syndrome caused by neoplasm is about 4 to 1.

Adenomas

Benign neoplasms usually weigh 10 to 50 g and most are less than 100 g. The tumors are distinctly encapsulated, lobulated, and yellow (Fig. 8.17). Rare focal areas of necrosis may be seen. Necrosis may be secondary to thrombosis from adrenal venography or from some other cause and is not necessarily a diagnostic feature of cortical carcinoma.

Microscopically, adenomas have clear cells that have abundant lipid and are arranged in sheets or nests (Figs. 8.18, 8.19). Nonneoplastic areas and the contralateral adrenal glands have lipid-depleted compact cells. Some lipid depleted cells may also be present in adenoma.

Ultrastructural features are similar to those of the normal adrenal cortex. Abundant smooth endoplasmic reticulum and tubulovesicular or tubulolamellar mitochondria are present in the cytoplasm. Osmophilic intracytoplasmic lipid and prominent microvilli can be seen in the neoplasm.

Carcinomas

Adrenal cortical carcinomas can range in weight from a few grams to about 5 k. Most carcinomas are larger than 100 g. The tumors are variably encapsulated, gray to brown, pink, or yellow with abundant necrosis and hemorrhage. Variable amounts of degenerative changes with coarse trabeculation may be seen. Some very large tumors may be quite cystic. Invasion or adhesion to adjacent structures is common. Careful dissection may reveal residual adrenal gland in some cases.

The histological appearance of cortical carcinomas can be quite variable. The pattern may vary from diffuse to alveolar or trabecular. Mixtures of lipid-poor and lipid-rich areas may be seen in adjacent areas. Some tumors may be remarkably uniform, while others may show marked pleomorphism with bizarre tumor cells. Extensive areas of hemorrhage, confluent necrosis, and mitotic figures including atypical mitoses are common (Fig. 8.20A and B). Vascular invasion may be seen but is less common and is diagnostic of malignancy when present.

Ultrastructurally, tumors causing Cushing's syndrome may have an abnormally large number of microvilli. The ultrastructural studies of Neville and others showed tumors with increased numbers of mitochondria and marked variation in organelle number in different cells (81,87).

Virilizing and Feminizing Adrenocortical Neoplasms

Neoplasms causing virilizing and feminizing syndromes often have similar histological features.

More than 80% of virilizing adrenocortical neoplasms are seen in females over a wide age range. Virilization may occur as part of Cushing's syndrome as well as independently of Cushing's syndrome.

Feminizing adrenal cortical tumors are very uncommon and most of these neoplasms are carcinomas. They occur most commonly during the fourth and fifth decades of life. In childhood they

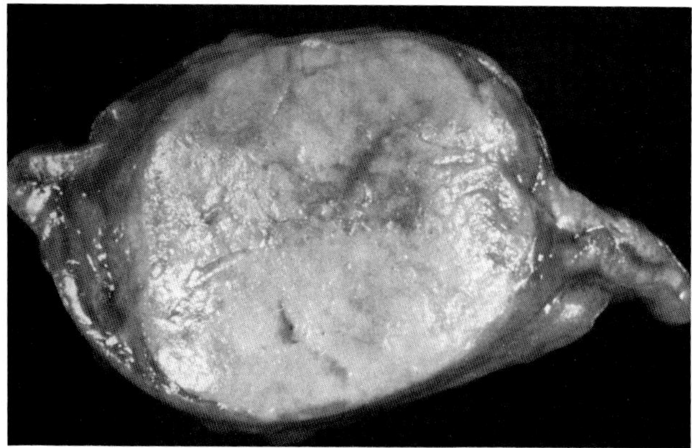

Figure 8.17. Adrenal cortical adenoma. The tumor is composed of yellow tissue and has a discrete capsule. Necrosis is not present.

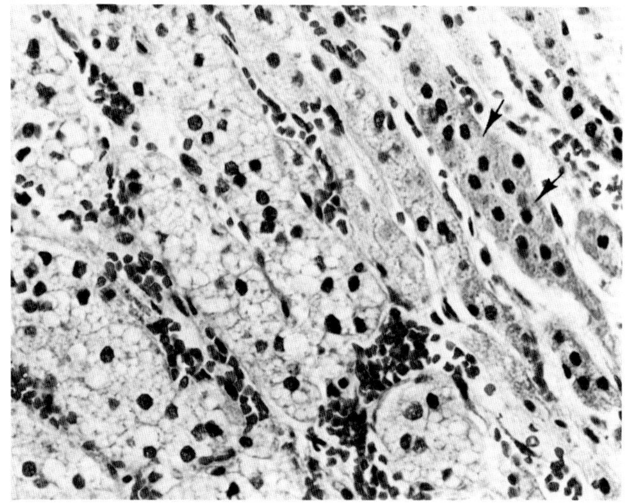

Figure 8.18. Microscopic sections of an adenoma from a patient with Cushing's syndrome. The tumor cells have abundant clear cytoplasm. At the periphery of the adenoma compressed cells with less cytoplasm are prominent (*arrows*).

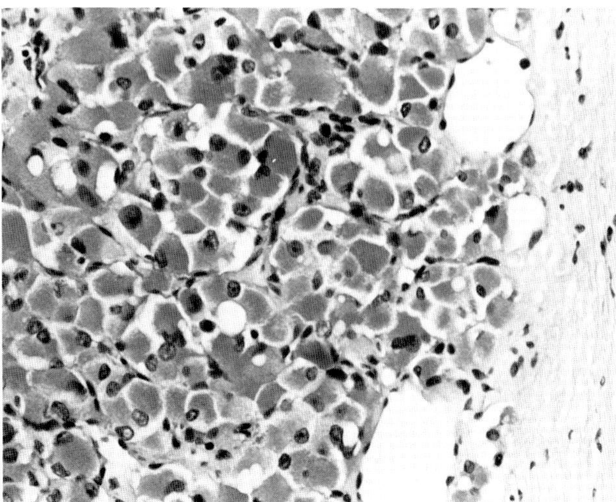

Figure 8.19. Adrenal cortical adenoma. Prominent cellular pleomorphism is present. This feature is common in cortical adenomas.

A

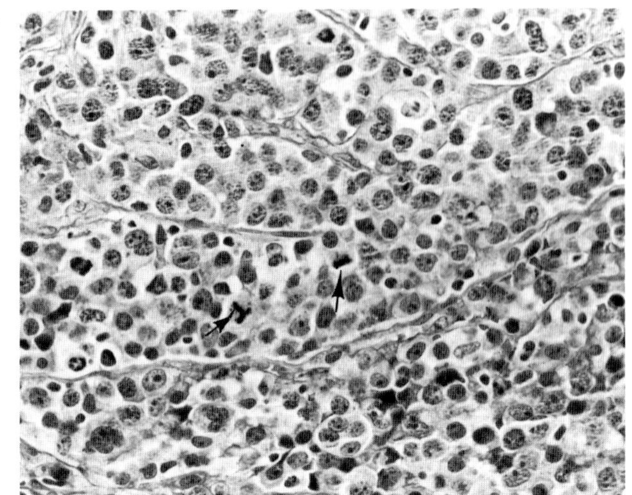

B

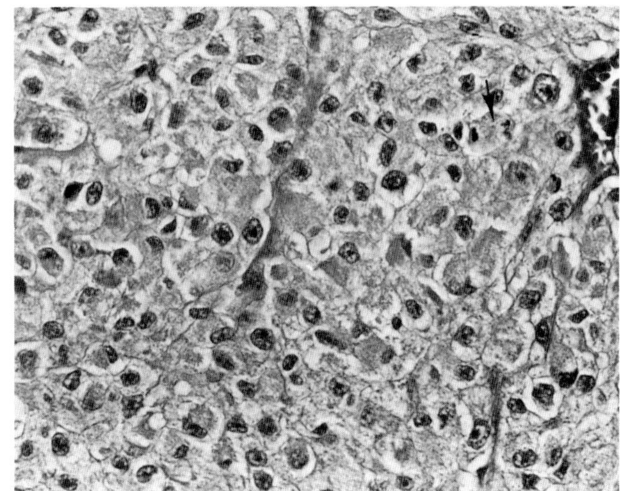

Figure 8.20. (A) Adrenal cortical carcinoma. Relatively uniform cells and many mitotic figures (*arrows*) including abnormal mitoses are present. (B) Carcinoma with prominent oncocytic features. Mitotic figures (*arrow*) are prominent. Areas of necrosis were also seen in the tumor.

can produce precocious puberty in girls or feminization in boys (45,117).

Adenomas

Benign neoplasms are usually encapsulated neoplasms that weigh between 50 and 300 g. The tumors are red-brown and 5 to 10 cm in diameter. Histologically, the cells are lipid poor and arranged in variable patterns including organoid, trabecular, or solid patterns. Some neoplasms may be quite pleomorphic with large nuclei and have very large cells with eosinophilic cytoplasm. Mitoses are absent or very rare.

Carcinomas

Carcinomas are large tumors and may weigh as much as 4 or 5 kg. The tumors are lobulated and range from pink-tan to yellow-gray. Areas of necrosis, hemorrhage, and cystic degeneration may be seen. Microscopically, the tumor cells range from uniform to markedly pleomorphic and multinucleated cells with numerous mitotic figures including atypical mitosis and areas of confluent necrosis.

Variants of Adrenal Cortical Neoplasms

Black adenomas are uncommon cortical neoplasms that have abundant lipofuscin pigment similar to the zona reticularis pigment. These tumors may be associated with hyperfunctional states including hyperaldosteronism or Cushing's syndrome or virilism (53,62).

Ectopic adrenal cortical tissues may occur in the liver, testis, kidney, gonads, and pancreas (109, 116). Ectopic adrenal cortical neoplasms can arise from these tissues in the liver, kidney, or in para-aortic regions (87). The histological features of these tumors may be similar to those arising in the adrenal glands.

Nonfunctioning Adrenal Cortical Neoplasms

About 10% of adrenal cortical neoplasms are clinically silent. Incidental solitary adrenal cortical nodules may be seen in 2 to 4% of autopsies. As previously discussed, some solitary nodules may not be true adenomas. The presence of a complete capsule with a single nodule usually favors an adenoma, whereas the cells of nonneoplastic nodules tend to merge with the adjacent adrenal cortical tissue. The presence of lipid-rich cells in an encapsulated lesion is usually diagnostic of an adenoma, although adrenal cortical nodules may also have lipid-rich cells.

Nonfunctional adrenal cortical carcinomas are uncommon (47,56). Like the functional adrenal cortical carcinomas, the tumors are large and have similar gross and histologic features including necrosis, hemorrhage, cystic changes, marked pleomorphism, eosinophilic and clear cells, mitotic activity, and confluent necrosis.

Biochemical and Cell Culture Studies of Adrenal Neoplasms

O'Hare and co-workers found that cultured adrenal cortical neoplastic cells produced different biochemical products that correlated to some degree with their biological behavior (86). Cells from functioning malignant tumors produced abnormal amounts of androgens and 11-deoxysteroids and had a blunted response to ACTH compared to adenomas. The utility of these studies as diagnostic aids remains to be determined.

In general, adrenal cortical neoplasms producing virilization or feminization are more likely to be malignant (98).

Morphological Distinction of Adrenal Cortical Adenomas and Carcinomas

Numerous studies have examined the morphological criteria that are helpful in distinguishing adenomas and carcinomas (44,105,121). The most helpful features that indicate malignancy include confluent tumor necrosis, tumor weight greater than 100 g, broad fibrous bands, a diffuse growth pattern, mitoses, especially atypical mitoses, and vascular invasion. No single criterion is absolute, since some benign tumors may weigh over 1000 g and have individual cell necrosis. Likewise, adrenal cortical carcinomas may weigh as little as 38 g (121). The presence of tumor giant cells and marked cellular pleomorphism are completely unreliable in distinguishing between cortical adenomas and carcinomas.

Metastasis of Adrenal Cortical Carcinomas

Adrenal cortical carcinomas are highly malignant neoplasms. About 25% of patients present with metastases. Common metastatic sites include liver, lungs, and lymph nodes. Less common sites are subcutaneous sites. Metastases to brain and bone are extremely uncommon (87). Metastatic carcinomas to the adrenal gland can be found in up to 27% of patients with carcinomas at autopsy (1). The most common primaries include lung, breast melanomas, and renal cell carcinomas.

Differential Diagnosis of Adrenal Cortical Neoplasms

Adenomas

The distinction between an adrenal nodule and an adenoma may be difficult. Nodules are often multiple, while adenomas are solitary and encapsulated. Atrophy and compression of the surrounding glands are present with functional adenomas but not nodules.

Distinction of an adenoma from a carcinoma has been previously discussed. Cortical adenomas may occasionally be difficult to distinguish from pheochromocytomas and metastatic carcinoma to the adrenal such as renal cell carcinoma. The presence of eosinophilic cytoplasmic inclusions in pheochromocytomas, as well as the use of immunohistochemical stains and ultrastructural studies, can help in these rare instances. Renal cell carcinomas can resemble cortical adenomas. Special stains for glycogen, which is abundant in renal cell carcinomas, and ultrastructural studies can help in this differential diagnosis.

Carcinomas

Renal cell carcinomas can be distinguished from adrenal cortical carcinomas by the criteria outlined above. Rare adrenal cortical tumors may occur ectopically in the kidney, and electron microscopic studies can help to confirm the diagnosis.

A few tumors, including hepatocellular carcinoma, alveolar soft part sarcoma, and large cell bronchogenic carcinoma, can look like cortical carcinoma on light microscopic examination. The presence of bile production and ultrastructural features can help to distinguish a hepatocellular carcinoma. Immunochemical staining for cytokeratin and alpha fetoprotein, which are commonly positive in hepatocellular carcinoma, can also be helpful. Alveolar soft part sarcomas may occur in the retroperitoneum occasionally. Diastase-resistant PAS positivity of the intracytoplasmic inclusions, ultrastructural features of smooth muscle differentiation, and immunohistochemical positivity for desmin can help to make the distinction.

Neuroendocrine tumors such as islet cell carcinomas, pheochromocytomas, and paragangliomas can be readily distinguished from adrenocortical carcinomas by ultrastructural studies to show secretory granules or by immunohistochemical staining for broad-spectrum markers such as chromogranins, neuron-specific enolase, and synaptophysin.

Bronchogenic carcinomas that are poorly differentiated may look histologically like cortical carcinomas, especially when they metastasize to the adrenals. Stains for mucin and keratin are extremely useful, as are ultrastructural studies to show features of steroid-producing cells or cells with tonofilaments and desomsomes, which are not present in adrenal cortical carcinomas.

Treatment

Surgical treatment is the principal therapy for adrenal cortical adenomas and carcinomas (38, 48). This is very effective for benign neoplasms.

Recurrences may be seen with incomplete tumor removal or if the operative field is contaminated with tumor fragments. Some patients with hyperaldosteronism may require bilateral excision of the adrenal to be cured. Rare benign tumors may be too large to resect completely and therapy with the adrenolytic agent 1,1-dichloro-2-(*o*-chlorophenyl)-2-(*p*-chlorophenyl)ethane (*o,p'*-DDD) or mitotane may be needed after surgery (Fig. 8.21).

Adrenocortical carcinomas may not be completely resectable in many cases, especially with very large tumors. Postoperative radiation therapy for residual tumor is often used. Chemotherapy with *o,p'*-DDD may alleviate symptoms and produce remission of tumor growth in some cases (4).

Mitotane has been used to treat adrenal cortical hyperplasia secondary to Cushing's disease. Chronic treatment can produce nodular atrophic adrenals. Microscopic examination shows fibrosis and enlarged and atrophic cells with nuclear pleomorphism.

Prognosis

Benign neoplasms have an excellent prognosis, except for local recurrence with tumors that are incompletely excised.

Carcinomas have a poor prognosis. Most patients develop metastases within 2 years of diagnosis and death usually occurs within 1 year after the development of metastases (46,85).

Other Tumors and Tumorlike Conditions

Adrenal Myelolipoma

These are benign tumors or tumorlike conditions composed of mature adipose tissue and bone marrow (69,88). They are usually asymptomatic and found as incidental lesions at autopsy or during

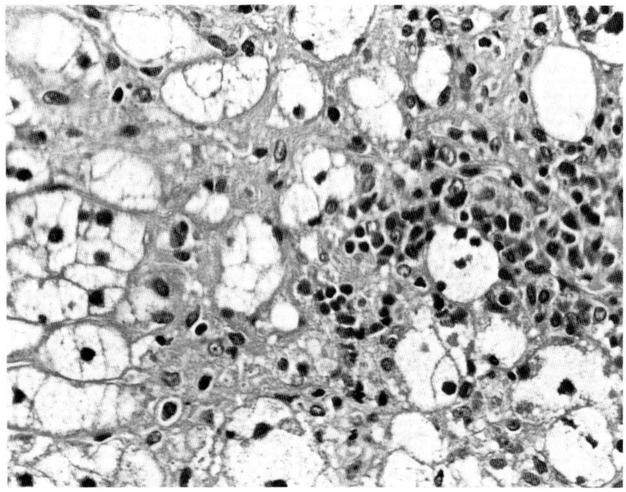

Figure 8.21. Adrenal cortical neoplasms treated with mitotane (*o,p'*-DDD). Areas of fibrosis and a lymphocytic infiltrate are present.

surgery in adults. Most lesions are 4 cm or smaller in diameter. Myelolipomatous changes may be found in diffuse adrenal cortical hyperplasia secondary to Cushing's disease and in bilateral pigmented nodular hyperplasia.

Grossly, the lesions are pale yellow with areas of pink or red from the hematopoietic elements. They are unencapsulated and may be lobulated. Occasionally these lesions are bilateral.

Microscopically, mature adipose tissues with hematopoietic elements are present. The hematopoietic tissues look like active bone marrow with various cell lines at different stages of maturation.

Lymphomas

With the exception of African Burkitt's lymphoma, which commonly invades the adrenal glands (24), primary and secondary lymphomas of the adrenals are rare (38a). Secondary lymphomas are more common than primary tumors. Plasmacytomas and diffuse and nodular lymphomas have all been reported to involve the adrenal glands.

Mesenchymal Neoplasms

Angiomas including hemangiomas and lymphangiomas have been observed in the adrenal cortex (93). Benign mesenchymal tumors such as neurofibromas, schwannomas, lipomas, and leiomyomas are seen occasionally. Rare malignant tumors such as leiomyosarcomas may arise in the adrenal gland, probably in association with the central vein (14). Carney has recently described a rare spindle cell lesion in the adrenal glands (11). The cells in this lesion may be similar to the wedge-shaped nodules in the adrenals that resembled ovarian stroma or thecal cells described by Fidler (29). These lesions, which are often bilateral in the few cases examined, may represent ovarian thecal metaplasia.

Adrenal Medulla

Adrenal Medullary Hyperplasia

Adrenal medullary hyperplasia is most commonly seen in patients with multiple endocrine neoplasia type 2a (II) and type 2b (III), but it may also be present with von Hipple-Lindau disease, in rare sporadic cases, in cases of sudden infant death syndrome probably secondary to chronic hypoxemia in cystic fibrosis, and in children and adults with hypertension (7,13,78,119).

Both of the adrenal glands are usually involved with medullary hyperplasia. The glands may be normal or slightly increased in size. Cut sections reveal expansion of medullary tissue into the tail region, which is abnormal because the tail of the gland does not normally have medullary tissue. Distinct nodularity may be seen grossly. Microscopically, the medullary cells may be increased in

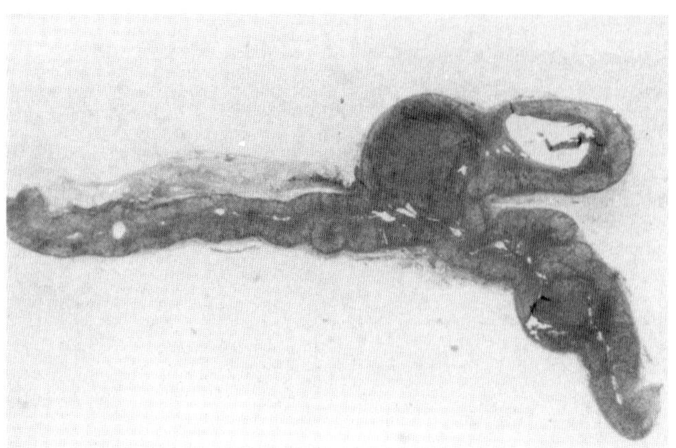

Figure 8.22. Adrenal nodular hyperplasia in a patient with multiple endocrine neoplasia type 2a.

size and have abundant eosinophilic to amphophilic cytoplasm. Mitotic figures are sometimes seen. Diffuse and nodular hyperplasia may be present (Fig. 8.22).

To establish the diagnosis of medullary hyperplasia it is necessary to perform morphometric analysis with a method such as point counting. With this technique the entire adrenal gland is necessary to assess the volume and weight of the cortex and medulla. DeLellis and his associates found that in the normal adrenal gland the corticomedullary weight was about 10 : 1 with an average medullary weight of 0.47 ± 0.15 g per gland. Patients with medullary hyperplasia have corticomedullary ratios of 5 : 1 with medullary weights of more than 1 g per gland (21).

Patients with familial adrenal medullary disease exhibit a spectrum of abnormalities ranging from diffuse and nodular hyperplasia to pheochromocytoma. Although it has been suggested that medullary nodules larger than 1.0 cm can be designated as pheochromocytomas (13), it is not known if nodules larger than 1.0 cm in patients with nodular hyperplasia represent true clonal neoplasms.

Sherwin suggested that there was a close relationship between adrenal medullary hyperplasia and development of pheochromocytoma (103). These observations were extended by the studies of Carney and associates (13) and DeLellis and associates (21). Patients with medullary thyroid carcinomas often have adrenal medullary hyperplasia early in the course of the disease. However, these individuals may go for a decade or longer before developing pheochromocytomas.

Neoplasms

Pheochromocytomas

Pheochromocytomas are rare tumors of the adrenal medulla that may occur at any age. They are most commonly seen between the third and fifth decades in sporadic cases. In familial cases the tumors are commonly diagnosed within the first two decades. In adults there is an equal sex ratio, but in children the male to female ratio is 2 : 1 (72,106). Familial cases are associated with various syndromes as summarized in Table 8.3.

Patients with pheochromocytomas have specific clinical signs and symptoms including hypertension, headaches, diarrhea, palpitation, pallor, and flushing. Myocardial damage from excessive production of catecholamines may develop (118). Some patients may have Cushing's syndrome associated with pheochromocytoma (73). Although historically many cases of pheochromocytomas were diagnosed at autopsy, this finding is relatively uncommon today because of earlier clinical diag-

Table 8.3. Familial disorders associated with pheochromocytoma

Syndrome	Bilateral disease	Nodular hyperplasia
MEN 2a (II)	Yes	Yes
MEN 2b (III)	Yes	Yes
Von Hippel-Lindau disease	Yes	Yes
Von Recklinghausen's disease (neurofibromatosis)	Rare	No
Sturge-Weber	Rare	No

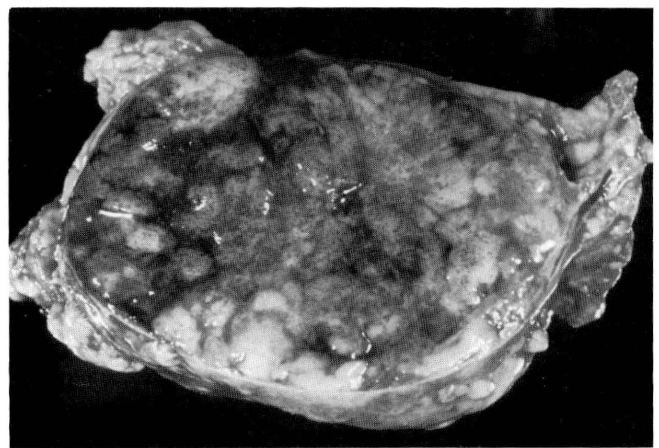

Figure 8.23. Pheochromocytoma. The tumor is tan and variably encapsulated.

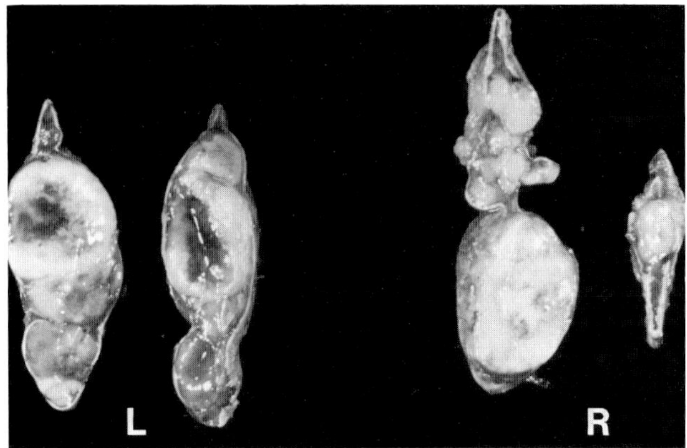

Figure 8.24. Pheochromocytoma in a patient with multiple endocrine neoplasia type 2b. Bilateral involvement of the adrenals and medullary hyperplasia are present.

nosis. Pheochromocytomas developing during pregnancy remain a significant cause of infant mortality (71).

The right adrenal is involved more commonly than the left in patients with sporadic pheochromocytomas. Most pheochromocytomas are gray-pink or tan and soft. Exposure to air leads to spontaneous oxidation of the endogenous catecholamines to adrenochrome and noradrenochrome pigments. The tumors are sometimes encapsulated. Larger tumors may have focal hemorrhage, necrosis, and cystic change and/or calcification (Fig. 8.23). Familial pheochromocytomas commonly involve both adrenal glands (Fig. 8.24).

Histologically, the tumor cells or pheochromocytes are arranged in cords or cell nests forming a zellballen pattern. Areas with trabecular and diffuse growth patterns are common. Occasional tumors have an alveolar pattern. Prominent capillary vascular structures are seen in the stroma. Cystic degeneration, focal hemorrhage, necrosis, or myxomatous change may be present. Rare tumors may have amyloid deposition. The tumor cells are generally polygonal and with granular cytoplasm that may be vacuolated (Fig. 8.25). PAS-positive eosinophilic droplets probably representing residual bodies are commonly present in the cytoplasm. Small foci of lymphocytes of unknown significance may be present in pheochromocytomas. Melicow (72) and others have described the prominent fetal fat (brown fat) present in the retroperitoneum of patients with pheocromocytomas. Prominent brown fat has also been observed in the absence of pheochromocytomas, so the significance of the association of pheochromocytomas and brown fat is unknown.

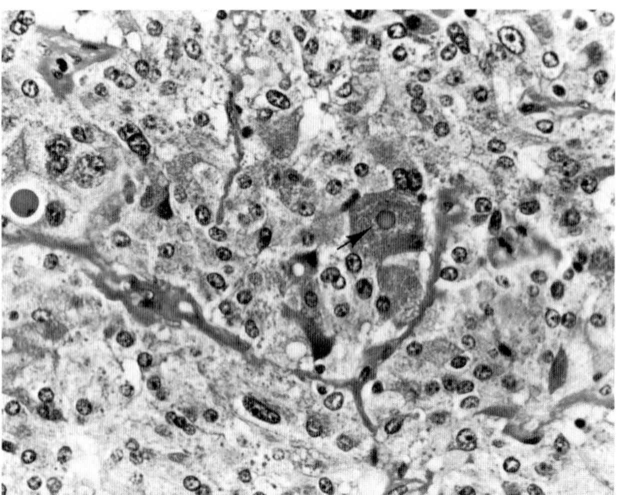

Figure 8.25. Histologic section of a pheochromocytoma showing pleomorphic cells with polygonal granular cytoplasm. Eosinophilic droplets (*arrow*) are present in the cytoplasm.

Ultrastructural features of pheochromocytomas fixed initially in glutaraldehyde show distinct norepinephrine-type granules with a halo between the dense granule core and the membrane and epinephrine-type granules with the dense core in close proximity to the granule membrane (Fig. 8.26). Mitochondria up to 3 μm in length have been described (120). Immunochemical staining for epinephrine and synaptophysin is usually positive in pheochromocytomas (Fig. 8.27A and B). In

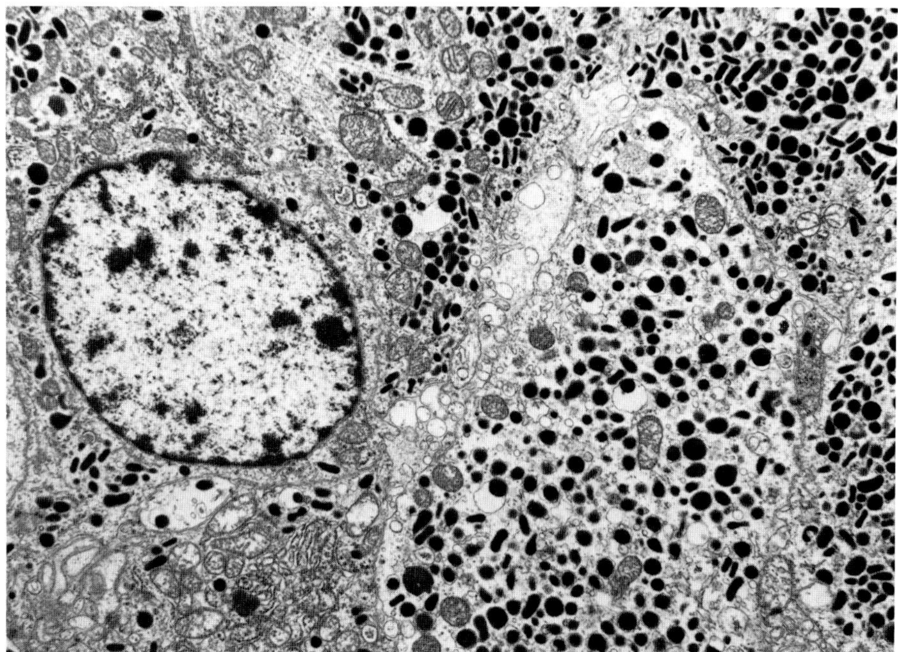

Figure 8.26. Ultrastructure of a pheochromocytoma. The tumor has many epinephrine-type granules with the dense core of the granule in close proximity to the membrane and a few norepinephrine-type granules with a distinct halo between the dense granule and the granule membrane (×6853).

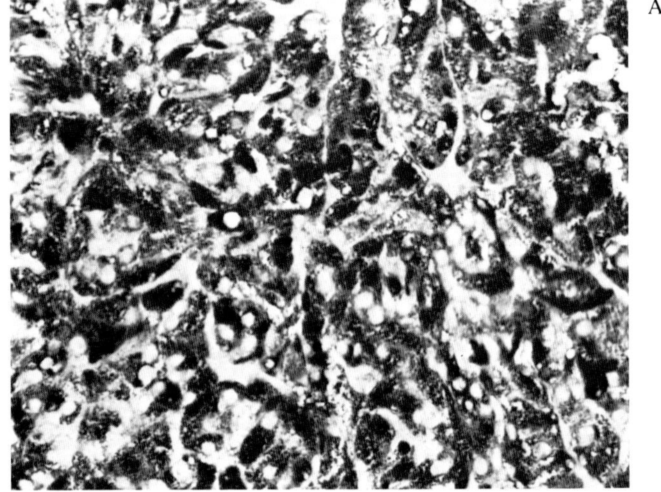

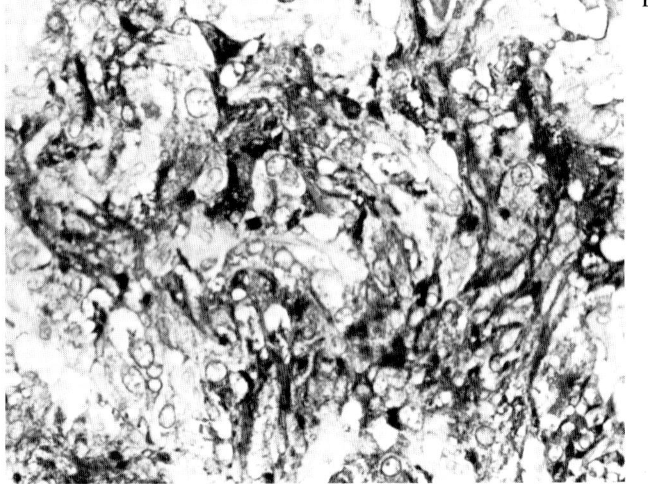

Figure 8.27. (A) Pheochromocytoma staining positive for epinephrine. (B) Synaptophysin immunoreactivity is seen in most of the tumor cells in this pheochromocytoma.

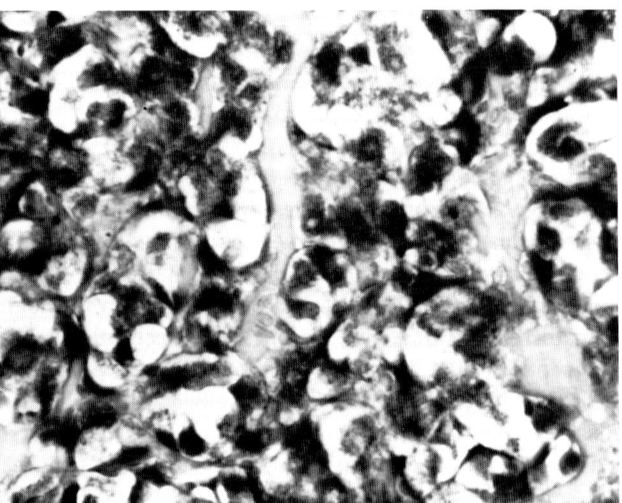

Figure 8.28. Chromogranin A messenger RNA in a pheochromocytoma detected with an oligonucleotide probe and a biotin system.

A

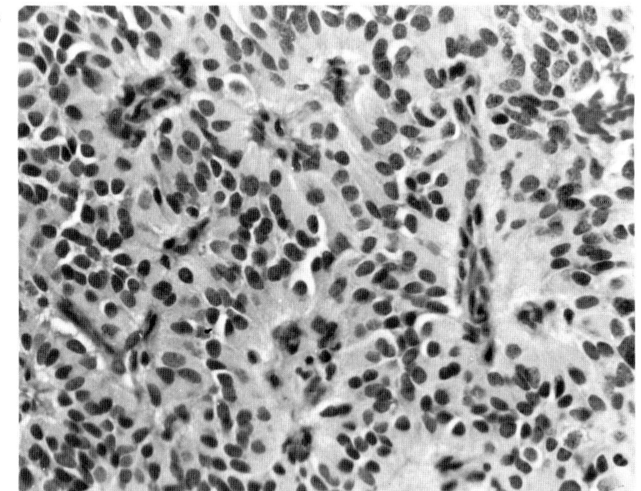

B

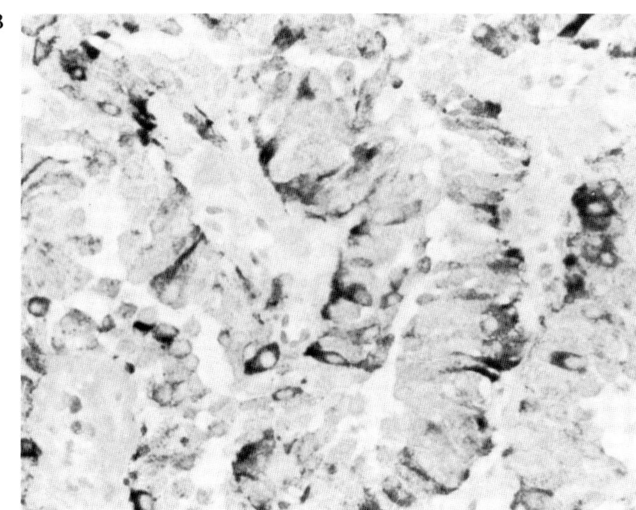

Figure 9.10. (A) Paraganglioma from spinal cord in the region of the cauda equina from a patient with leg weakness. The tumor has many central vessels and looks like a ependymoma. Staining for chromogranin A was positive (B).

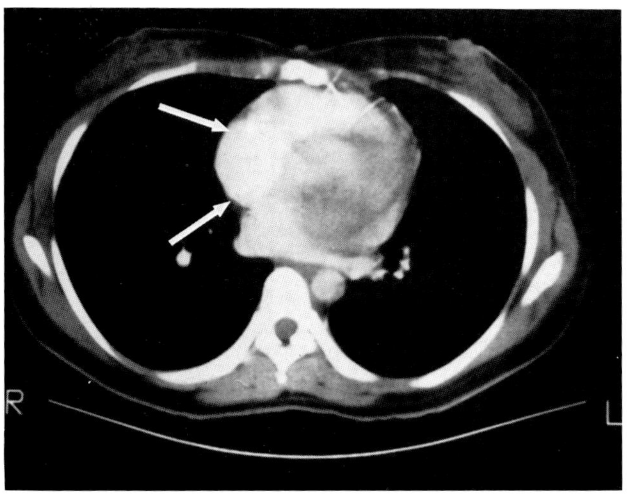

Figure 9.11. A CT scan of a cardiac paraganglioma shows a large tumor in the pericardium involving the atrium (*arrows*).

Figure 9.12. Cardiac paraganglioma. The tumor has focal cystic areas and is tan-yellow.

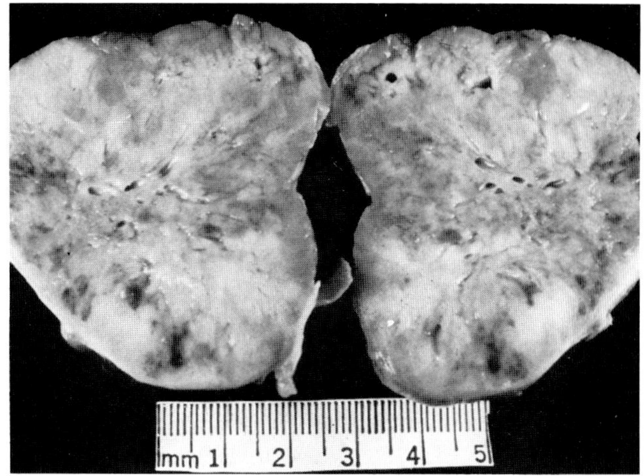

Figure 9.13. Histologic section of a cardiac paraganglioma showing a prominent zellballen pattern and slight nuclear pleomorphism.

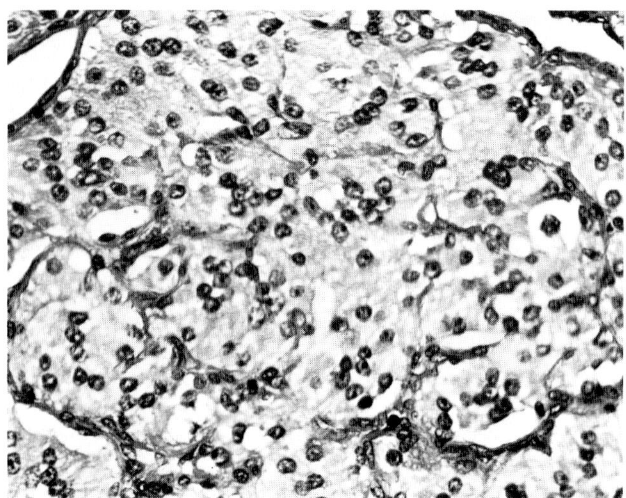

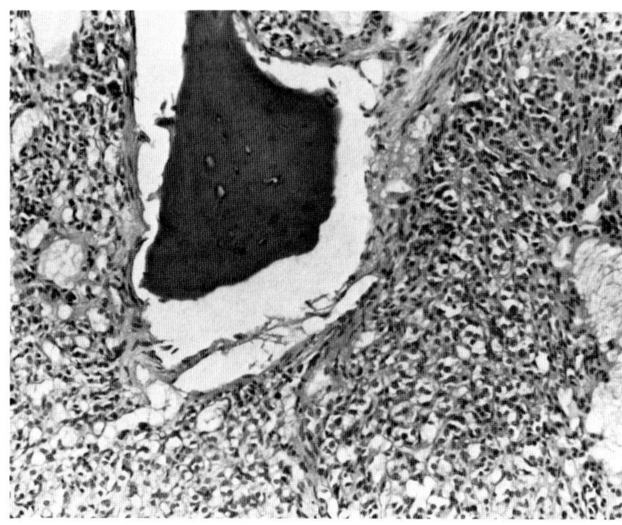

Figure 9.14. Malignant paraganglioma from the carotid body. The tumor metastasized to the vertebrae. The histological appearance of this tumor was quite bland.

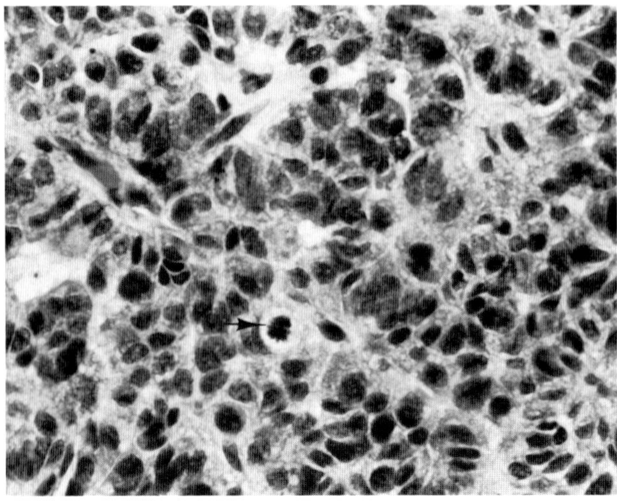

Figure 9.15. Malignant paraganglioma from the organ of Zuckerkandl metastatic to the liver. This tumor had a high mitotic rate (*arrow*).

leiomyosarcomas and pulmonary chondromas by Carney and his associates (3).

Differential Diagnosis

The differential diagnosis of paragangliomas includes alveolar soft part sarcomas, hemangiomas, glomus tumors, olfactory neuroblastomas, medullary thyroid carcinomas, and other neuroendocrine neoplasms. PAS-positive, diastase-resistant globules and variable positive immunoreactivity for desmin with negative staining for chromogranin can separate paragangliomas from alveolar soft part sarcomas (Fig. 9.16). Jugulotympanic and other paragangliomas may be confused with hemangiomas. The zellballen pattern and prominent immunoreactivity for broad-spectrum endocrine markers can help to distinguish these lesions. Some glomus tumors of soft tissues and stomach may be difficult to distinguish from paragangliomas, although immunostaining readily solves this problem. Olfactory neuroblastomas often have neurofibrillary material, rosettes, and a less prominent zellballen pattern than paragangliomas. Medullary thyroid carcinomas can look very much like paragangliomas. The presence of amyloid and positive immunostaining for calcitonin can be helpful in separating paragangliomas of the neck from medullary thyroid carcinomas.

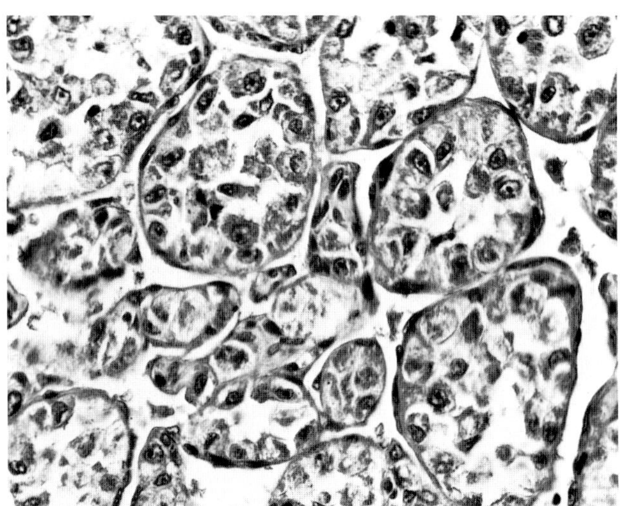

Figure 9.16. Alveolar soft part sarcoma. The organoid or nestlike appearance simulates a paraganglioma.

Treatment

Surgery is the treatment of choice for most paragangliomas. Incomplete excision such as with jugulotympanic paraganglia is associated with recurrences. Radiation therapy is often only palliative in the treatment of paragangliomas (6,10,36). Metaiodobenzylguanidine labeled with radioactive iodine has been used in the localization and in the treatment of some malignant paragangliomas (36,37a).

Prognosis

The prognosis for benign paragangliomas that are completely excised is excellent. Incompletely excised tumors such as many jugulotympanic paragangliomas have frequent recurrences.

Paragangliomas of the anterior mediastinum often have an aggressive biological course because of the difficulties in resection, and these tumors may be associated with a high fatality rate. Paragangliomas of the organ of Zuckerkandl have a high metastatic rate (7,25,33,44).

Malignant paragangliomas may be associated with metastasis to bone, lymph nodes, liver, lung, and dura and can lead to considerable morbidity and increased mortality.

References

1. Arias-Stella J, Valcarcel J (1976) Chief cell hyperplasia in the human carotid body at high altitudes. Physiologic and pathologic significance. Hum Pathol 7:361–373
2. Berdal P, Braaten M, Cappelen C Jr, Mylius EA, Walaas O (1962) Noradrenaline-adrenaline producing nonchromaffin paraganglioma. Acta Med Scand 172:249–257
3. Carney JA, Sheps SG, Go VLW, Gordon H (1977) The triad of gastric leiomyosarcoma, functioning extra-adrenal paraganglioma and pulmonary chondroma. N Engl J Med 296:1517–1518
4. Chase WH (1933) Familial and bilateral tumours of the carotid body. J Pathol Bacteriol 36:1–12
5. Cragg RW (1934) Concurrent tumors of the left carotid body and both Zuckerkandl bodies. Arch Pathol 18:635–645
6. Drasin H (1978) Treatment of malignant pheochromocytoma. West J Med 128:106–111
7. Enzinger FM, Weiss SW (1988) Soft tissue tumors. St. Louis, Mosby, pp 836–860
8. Gallivan MVE, Chun B, Rowden G, Lack EE (1980) Intrathoracic paravertebral malignant paraganglioma. Arch Pathol Lab Med 104:46–51
9. Gallivan MVE, Chun B, Rowden G, Lack EE (1979) Laryngeal paraglioma: a case report with ultrastructural analysis and literature review. Am J Surg Pathol 3:85–92
10. Gaylis H, Mieny CJ (1977) The incidence of malignancy in carotid body tumours. Br J Surg 64:885–889
11. Glenner GG, Crout JR, Roberts WC (1961) A noradrenaline secreting carotid body-like tumour. Lancet 2:439
12. Glenner GG, Crout JR, Roberts WC (1961) Paragangliomas of the head and neck region. A pathologic study of tumors from 71 patients. Hum Pathol 10: 191–218
13. Glenner GG, Grimley PM (1974) Tumors of the Extra-Adrenal Paraganglion System (Including Chemoreceptors). Atlas of Tumor Pathology, Second Series, Fascicle 9. Washington, DC, Armed Forces Institute of Pathology
14. Glenn F, Gray GF (1976) Functional tumors of the organ of Zuckerkandl. Ann Surg 183:578–586
15. Gould VE, Sommers SC (1982) The adrenal medulla and paraganglia. In: Bloodworth JMB Jr (ed) Endocrine Pathology, General and Surgical. Baltimore, Williams & Wilkins, pp 497–511
16. Hatfield PM, James AE, Schulz MD (1972) Chemodectomas of the glomus jugulare. Cancer 30: 1164–1168
17. Heath D, Edwards C, Harris P (1970) Post mortem size and structure of the human carotid body. Its relation to pulmonary disease and cardiac hypertrophy. Thorax 25:129–140
18. Heppeston AG (1958) A carotid body-like tumour in the lung. J Pathol 75:461–464
19. Jago R, Smith P, Heath D (1984) Electron microscopy of carotid body hyperplasia. Arch Pathol Lab Med 108:717–722
20. Johnson TL, Lloyd RV, Shapiro B, Sisson JC, Beierwaltes WH, Thompson NW, Orringer MB (1985) Cardiac paragangliomas. A clinicopathologic and immunohistochemical study of four cases. Am J Surg Pathol 9:827–834
21. Johnson TL, Zarbo RJ, Lloyd RV, Crissman JD (1988) Paragangliomas of the head and neck: immunohistochemical neuroendocrine and intermediate filament typing. Mod Pathol 1:216–223
22. Korn D, Bensch K, Liebow AA, Catleman B (1960) Multiple minute pulmonary tumors resembling chemodectomas. Am J Pathol 37:641–672
23. Lack EE (1977) Carotid body hypertrophy in patients with cystic fibrosis and cyanotic congenital heart disease. Hum Pathol 8:39–51

tinction of Merkel cell carcinomas from metastatic neuroendocrine carcinoma requires careful clinopathological correlation. The finding of "fibrous bodies" or juxtanuclear collections of intermediate filaments with associated dense-core secretory granules is relatively uncommon in other neuroendocrine carcinomas, albeit this feature is not unique in Merkel cell carcinoma.

Treatment and Prognosis

Surgical excision is the treatment of choice for these neoplasms. Radiation therapy and chemotherapy have been used for recurrent disease. Combination chemotherapy has led to complete response in very few patients with metastatic disease (21).

Metastatic disease usually involves regional lymph nodes most commonly, while distant metastasis involves skin, subcutaneous tissue, liver, bone, lungs, pancreas, pelvis, and retroperitoneal soft tissues (55). Regional lymph node metastases occur in more than half of all patients (56%), while distant metastasis occur in about 28%. Local recurrence occurs in about one-third of all patients with disease and about one-third of patients with neuroendocrine carcinomas die of their disease.

Larynx

Neuroendocrine carcinomas of the larynx are uncommon tumors (26,76). About 94 cases have been reported in the literature (5,68a,76). These tumors occur in adults with an age range of 37 to 83 years. The supraglottic larynx is the most common site and a history of cigarette smoking is common in patients with these neoplasms.

Pathology

The tumors range from small polypoid to large bulging masses with a gray-white or tan cut surface. Histologically, these tumors can be divided into two groups (76). The small-cell neuroendocrine carcinoma is similar to these tumors in the lung with small amounts of cytoplasm, small nuclei, and coarse chromatin. The large-cell carcinomas have moderate to abundant cytoplasm and can be arranged in nests, cords, ribbons, or acini (Fig. 10.4). Ultrastructural studies reveal small secretory granules 70 to 250 nm in diameter (Fig. 10.5). Immunohistochemical staining is positive for chromogranin A and for neuron-specific enolase. Chromogranin A is present focally in the small-cell variant of laryngeal neuroendocrine carcinoma. Calcitonin, somatostatin, and serotonin immunoreactivity may also be present. Woodruff and his co-workers (76) found calcitonin immunoreactivity in 67% of the large-cell variant of laryngeal neuroendocrine carcinomas.

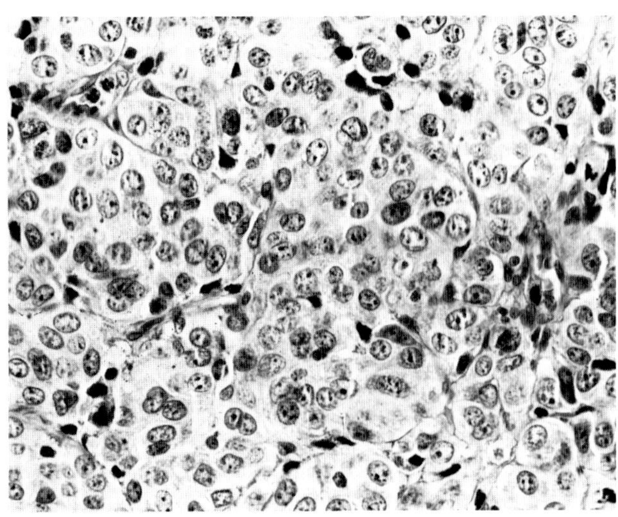

Figure 10.4. Laryngeal neuroendocrine carcinoma of the large-cell type. The tumor cells have abundant cytoplasm and large nuclei and are growing in a slightly organoid pattern.

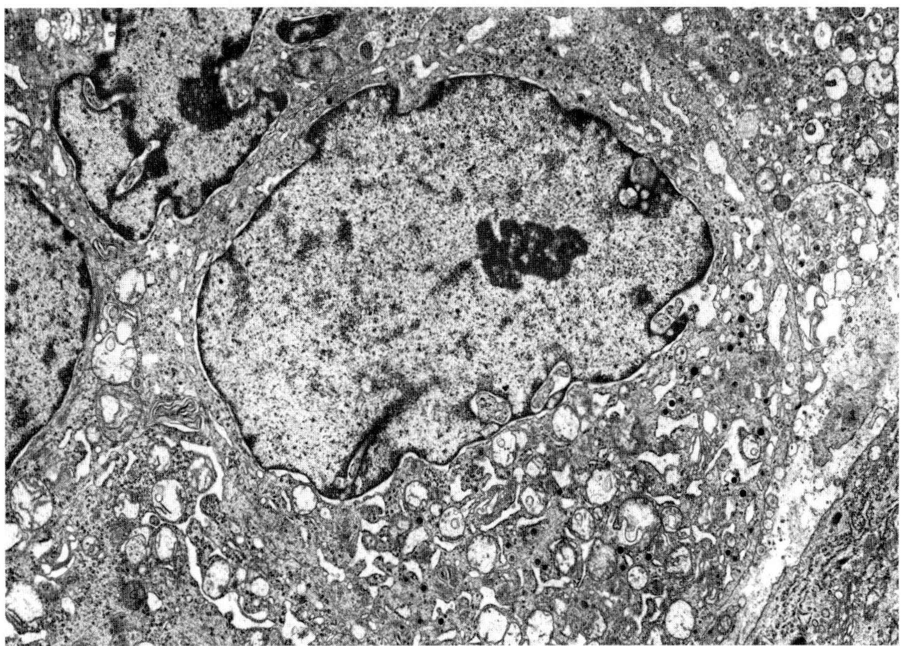

Figure 10.5. Electron micrograph of a laryngeal neuroendocrine carcinoma. Many small secretory granules 70 to 250 nm in diameter are present (×6450).

Differential Diagnosis

The differential diagnosis includes lymphoma, squamous cell carcinoma, medullary thyroid carcinoma, paraganglioma, and metastatic neuroendocrine carcinoma. Immunohistochemistry and electron microscopy can readily separate neuroendocrine carcinomas from lymphomas and squamous cell carcinomas. Paragangliomas usually have a prominent zellballen pattern and are usually negative for calcitonin on immmunostaining. The secretory granules of paragangliomas are often larger than those of laryngeal neuroendocrine carcinomas. Because the large-cell variant of laryngeal neuroendocrine carcinoma is often positive for calcitonin, it is often difficult to separate these lesions from medullary thyroid carcinomas. Amyloid can be found in both tumors, so clinicopathological studies are very important to make this distinction. In most cases of medullary thyroid carcinoma the serum calcitonin is elevated, while elevated serum calcitonin is rare in large-cell laryngeal neuroendocrine carcinoma.

Treatment and Prognosis

Surgical treatment is the therapy of choice. Local recurrences are seen in about a third of cases after surgery (19). Radiation therapy and chemotherapy usually do not produce an objective response (5).

Cervical lymph nodes are the most common site of metastatic disease. Metastatic foci to the skin, lungs, liver, heart, and bone have also been reported.

In the study of Woodruff and co-workers, patients with small-cell carcinoma survived a mean of 11 months and those with large-cell carcinoma a mean of 36 months. These findings with small-cell neuroendocrine carcinoma of the larynx are in agreement with studies of others that show the small-cell carcinoma to be very aggressive.

Breast

Although cells with neuroendocrine differentiation can be found in breast carcinomas, neuroendocrine carcinomas of the breast are very rare. Some investigations have found argyrophilic cells staining for chromogranin in the normal breast (9,10).

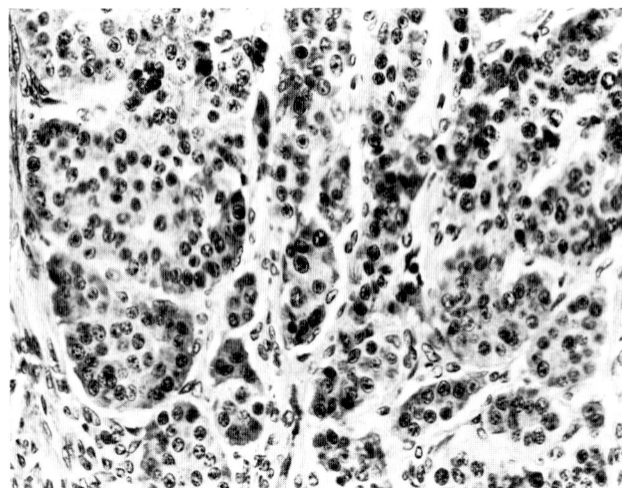

Figure 10.7. Histologic section of a carcinoid tumor of the ovary. The tumor cells show a predominantly insular and focal trabecular growth pattern.

A

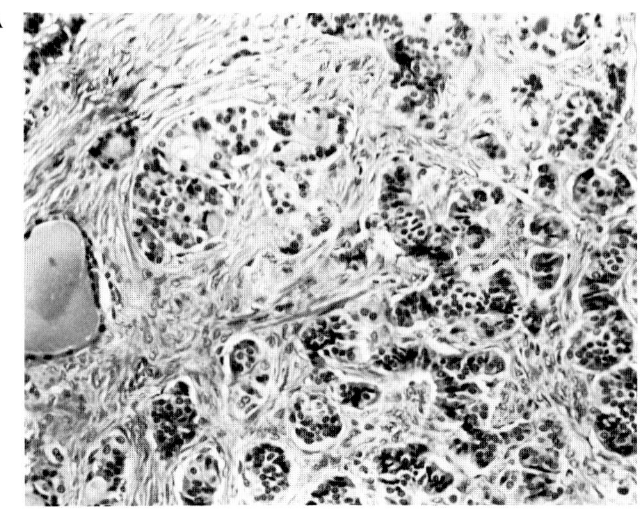

B

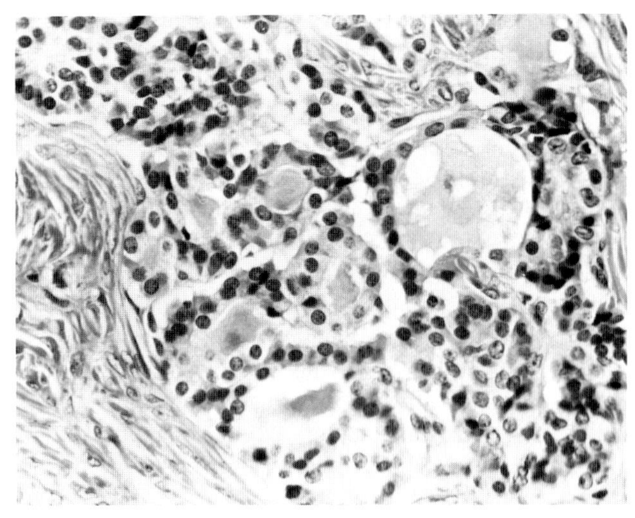

Figure 10.8. (A) Strumal carcinoid tumor. A prominent trabecular pattern and colloid follicles that contain thyroglobulin are present. (B) Higher magnification of a strumal carcinoid tumor. Thyroglobulin-containing follicles are prominent. Some of the carcinoid tumor cells express calcitonin hormone and serotonin.

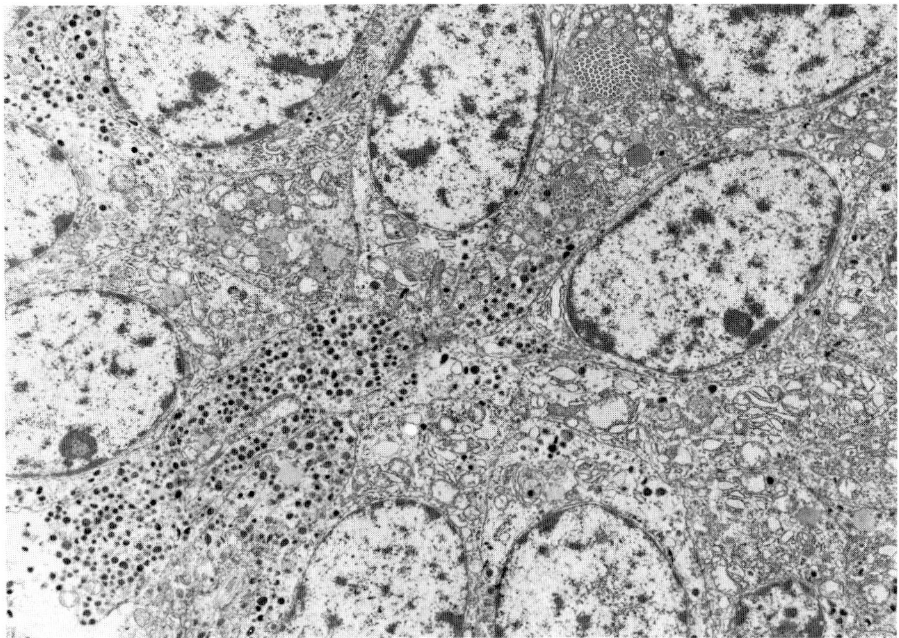

Figure 10.9. Electron micrograph of an ovarian carcinoid. The tumor, which had a trabecular growth pattern, contains round to oval secretory granules 150 to 400 nm in diameter (×4028).

Pathology

Insular carcinoids are usually unilateral, although the other ovary may contain a dermoid cyst. The tumor cells form islands with acini within the islands (Fig. 10.7). The tumor cells are often argentaffin positive and usually argyrophilic. Ultrastructural examination shows dense-core granules that are often dumbbell shaped.

Trabecular carcinoids consist of ribbons of tumor cells with a fibrous stroma. The cells may be argentaffin positive and are usually argyrophilic. Ultrastructural examination reveals round to oval granules (Fig. 10.7).

Strumal carcinoids can have a trabecular, insular, or mixed pattern along with the thyroid tissue (Fig. 10.8a and B). Some tumors are argentaffin positive, and most are argyrophilic. Ultrastructural studies reveal round to oval secretory granules 100 to 400 nm in diameter (Fig. 10.9). A variety of neuropeptides and other endocrine markers have been described in ovarian carcinoids, including chromogranin, thyroglobulin, calcitonin, serotonin, somatostatin, glucagon, insulin, and gastrin (40,46,58,59,64). Calcitonin and thyroglobulin immunoreactivity has been found in strumal carcinoids, but amyloid has not been found in these neoplasms (46).

Differential Diagnosis

The differential diagnosis of insular and trabecular carcinoid tumors includes granulosa cell tumor, Sertoli-Leydig cell tumor, and metastatic neuroendocrine carcinoma. Granulosa cell tumors have nuclear grooves, scant cytoplasm, and Call-Exner bodies. Sertoli-Leydig cell tumors have features of steroid-producing cells, including smooth endoplasmic reticulum and absence of neuroendocrine markers or secretory granules. Metastatic neuroendocrine carcinomas must be excluded by extensive clinicopatholoic evaluation.

Treatment and Prognosis

Surgical treatment is the therapy of choice for ovarian carcinoid tumors. These tumors are low-grade neoplasms with only a limited potential for metastasis (52). Rare cases of recurrent disease in the

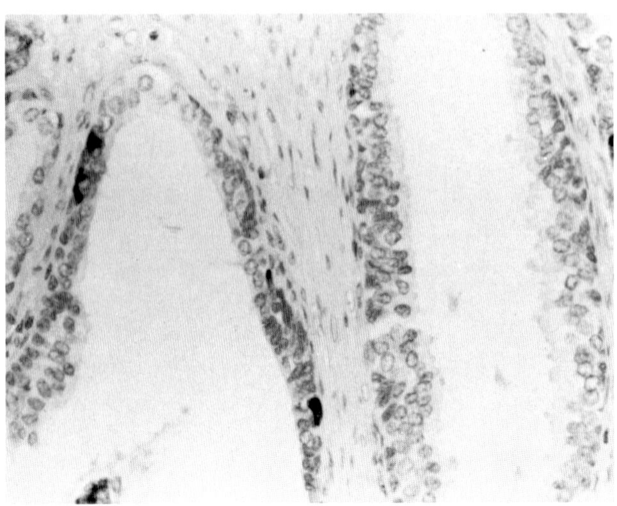

Figure 10.10. Chromogranin A immunoreactivity in the endocrine cells of normal prostatic epithelium. The endocrine cells are located in the basal part of the epithelium away from the lumen.

peritoneal cavity have been seen with insular carcinoids associated with the carcinoid syndrome (52).

Prostate

Neuroendocrine cells have been recognized in the normal prostate by argyrophilic histochemical staining and immunohistochemical techniques (Fig. 10.10). di Sant'Agnese and co-workers found serotonin, calcitonin, and somatostatinlike immunoreactivities in normal prostatic glands (17,18). Prostatic carcinomas with neuroendocrine features can be part of adenocarcinomas or can be pure small-cell carcinomas of the prostate (1,20,25,49, 63,65,69). Most studies show that the neoplasms with neuroendocrine differentiation are quite heterogeneous. Immunohistochemical studies with broad-spectrum endocrine markers and with antibodies to various peptides reveal ACTH, somatostatin, glucagon, and calcitonin immunoreactivities. Markers for prostatic epithelial differentiation, including prostate-specific antigen and prostatic acid phosphatase, are usually present (1,63,65). Some patients may have Cushing's syndrome, usually due to ACTH production (25,65,69).

Patients with small-cell carcinoma of the prostate are usually in the same age range as those with prostatic adenocarcinoma. In the recent study by Tetu and co-workers, the 20 patients ranged from 30 to 89 years with a median of 67 years and 2 patients were younger than 50 years (63).

Pathology

Histologically, the tumors are usually admixed with adenocarcinomas. The small-cell carcinomas can range from oat cell type to intermediate cell types with focal insular or trabecular patterns. Argyrophilic stains are often positive in small-cell carcinomas and ultrastructural studies revealed dense-core granules up to 250 nm in diameter in some tumors (63). Immunohistochemical stains reveal many peptide hormones, often in a patchy distribution.

Differential Diagnosis

The differential diagnosis includes poorly differentiated prostatic carcinomas and metastatic endocrine tumors to the prostate. Immunohistochemical staining and clinicopathological studies can often help to confirm the diagnosis and to exclude prostatic carcinomas or metastatic neuroendocrine carcinoma (Fig. 10.11A, B).

Treatment and Prognosis

Surgical treatment is often done for these neoplasms. The prognosis of patients with small-cell carcinoma is poor, with most patients dying of their disease less than 2 years after diagnosis. In one study, patients who presented with a pure adenocarcinoma and subsequently developed a small-cell component generally did better than

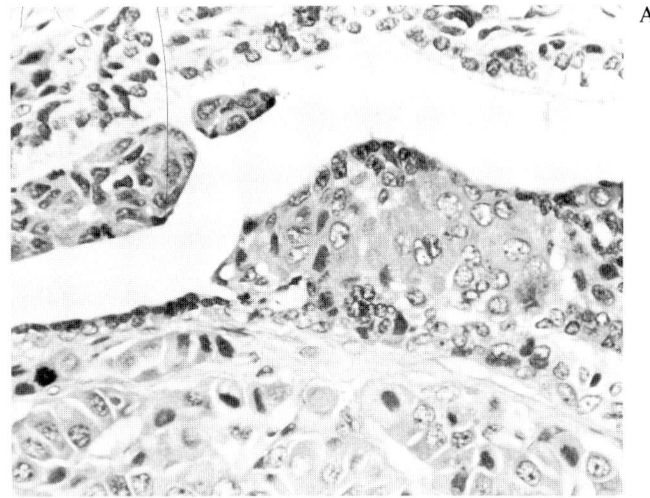

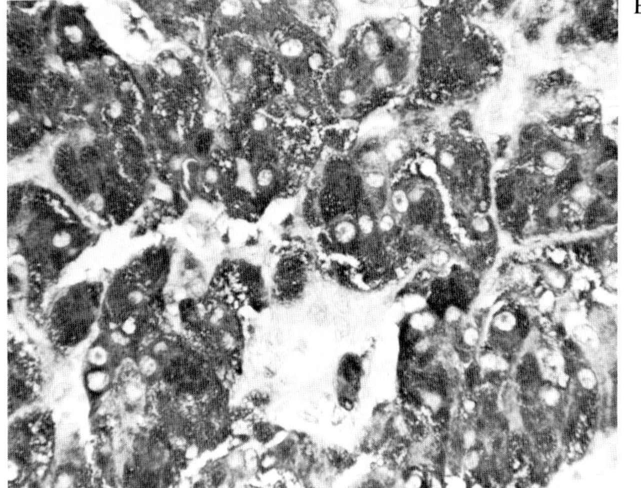

Figure 10.11. (A) Unusual case of a medullary thyroid carcinoma metastatic to prostate in a young man with multiple endocrine neoplasia type 2b. The tumor cells have abundant eosinophilic cytoplasm. (B) Immunostaining for calcitonin was diffusely positive in the tumor cells. Stains for chromogranin A were also positive. The tumor cells were negative for somatostatin, prostate-specific antigen, and prostatic acid phosphatase.

those with a small-cell component at the time of diagnosis (63).

Testicular Carcinoids

Primary testicular carcinoids are rare neoplasms. Like ovarian carcinoids, they can be part of a teratoma or be a pure testicular carcinoid. Pure carcinoid tumors are more common. An irregular pattern similar to that of midgut carcinoids is most common (6,44). Carcinoid syndrome is rarely seen in spite of the fact that the venous drainage of the testes bypasses the liver, going directly into the systemic circulation (32).

Pathology

The tumor cells are polygonal with round central nuclei and moderate amounts of eosinophilic cytoplasm. Argyrophilic stains are usually positive and some tumors are also argentaffin positive.

Ultrastructural studies often reveal dense-core granules ranging from 70 to 300 nm in diameter. Reported immunochemical stains in the study revealed positive staining for vasoactive intestinal polypeptide, substance P, and serotonin (44).

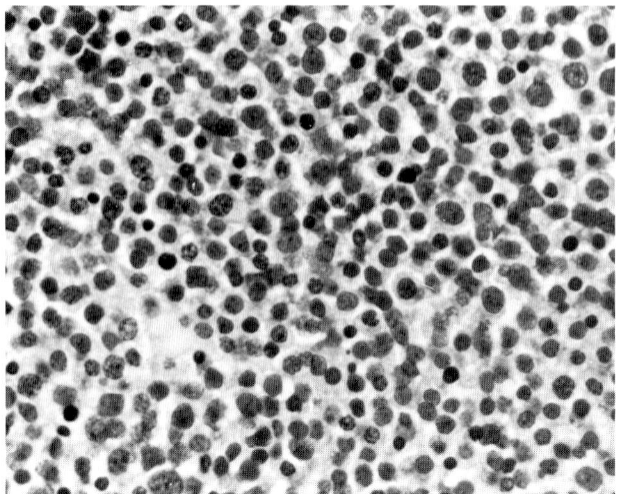

Figure 10.12. Olfactory neuroblastoma. The tumor is composed of sheets of small cells with a slightly fibrillary background.

Differential Diagnosis

The differential diagnosis includes malignant lymphomas, melanomas, Leydig cell tumors, neuroblastomas, and metastatic neuroendocrine carcinomas. Immunohistochemical stains for broad-spectrum neuroendocrine markers and for hematopoietic markers can exclude lymphomas. Negative staining for S100 and HMB-45 can exclude melanomas. Ultrastructural examination for secretory granules and the absence of Reinke crystals can help to differentiate Leydig cell tumors. Neuroblastomas and metastatic neuroendocrine carcinomas can be excluded only with clinicopathologic correlation along with ultrastructural studies. Neuroblastomas often have rosettes and neurofibrillary material, while at the ultrastructural level small secretory granules and microtubules can be helpful differential features.

Testicular carcinoids should be considered as low-grade malignant neoplasms. In a few cases metastatic disease has been reported (6,61).

Other Carcinoid Tumors

Occasionally, primary carcinoid tumors may be seen in the kidney (60), paranasal sinus (35), larynx (8), liver (27), gallbladder (66), middle ear (41), eye and orbit (45), and mesentery (3). Some of these neoplasms represent variants of other neuroendocrine tumors rather than true carcinoid tumors like the ones in the gastrointestinal tract and lungs.

Olfactory Neuroblastoma

The olfactory neuroblastoma or esthesioneuroblastoma is a malignant neoplasm with neuroendocrine differentiation that arises from the olfactory mucosa in the nasal cavity. Some of these neoplasms have been shown to produce catecholamines and vasopressin (38,57). These tumors can be seen in children and adults with a bimodal peak in the second and sixth decades (33).

Pathology

Grossly, the tumors appear as polypoid masses in the nasal cavity or ethmoid sinus. The tumors are friable and bleed easily.

Histologically, the tumor is composed of small cells with a diffuse growth pattern or with irregular nests of cells (Fig. 10.12). There is usually slight to moderate nuclear pleomorphism and variable numbers of mitotic figures. A fibrillary background is often present but may be sparse. Flexner or Homer-Wright rosettes may be seen in some tumors. In rare instances ganglion cells may be present.

Immunohistochemical staining for neuron-specific enolase, chromogranins, and other broad-spectrum neuroendocrine markers may be helpful in less well differentiated tumors (62). Ultrastructural studies often show small secretory granules.

Differential Diagnosis

The differential diagnosis includes rhabdomyosarcomas, lymphomas, anaplastic carcinomas, Ewing's sarcoma, pituitary adenomas, ependymomas, and other neuroendocrine tumors. Immunohistochemical and ultrastructural studies can help to distinguish between these tumors in most cases.

Treatment and Prognosis

Olfactory neuroblastomas often invade locally, but some tumors may metastasize to regional lymph nodes, lungs, and bone (33).

Surgical treatment and radiation therapy are the usual treatment (2), while chemotherapy is used for metastatic disease.

A 5-year survival of 50% has been reported, and the worst prognosis is seen in patients with higher-grade tumors (2).

Multidirectional Differentiation in Neoplasms

Neoplasms with endocrine and nonendocrine populations are found in tissues such as the prostate (25,63,65), breast (13,42,43,75), endometrium (34,53), and other sites (11,16). Many examples of multidirectional differentiation of neoplasms have been discussed in earlier chapters, including goblet cell carcinoid tumors of the appendix with mucin production (48) and mixed follicular-medullary thyroid carcinomas of the thyroid producing both calcitonin and thyroglobulin (30,31). The existence of such neoplasms raises many questions about histogenesis and microenvironmental influences on the development of specific neoplasms. DeLellis and his co-workers have suggested that tumors with multidirectional differentiation may arise from uncommitted stem cells, which normally may have the capacity to differentiate into multiple different cell types (16). Experimental evidence to support this concept has been presented from cell culture and other developmental studies (7,14).

References

1. Abrahamsson PA, Wadtrom LB, Alumets J, Falkmer S, Grimelius L (1987) Peptide hormone- and serotonin-immunoreactive tumour cells in carcinoma of the prostate. Pathol Res Pract 182:298–307
2. Baker DC, Perzin KH, Conley J (1979) Olfactory neuroblastoma. Otolaryngol Head Neck Surg 87: 279–283
3. Barnardo DE, Sravrou M, Bourne R, Bogomoletz WV (1984) Primary carcinoid tumor of the mesentery. Hum Pathol 15:796–798
4. Barrett RJ, Davos I, Leuchter RS, Lagasse LD (1987) Neuroendocrine features in poorly differentiated and undifferentiated carcinomas of the cervix. Cancer 60:2325–2330
5. Baugh RF, Wolfe GT, Lloyd RV, McClatchey KD, Evans DA (1987) Carcinoid (neuroendocrine carcinoma) of the larynx. Ann Otol Rhinol Laryngol 96: 315–321
6. Berdjis C, Mostofi FK (1977) Carcinoid tumors of the testis. J Urol 118:777–782
7. Bjerknes M, Cheng H (1981) The stem-cell zone of the small intestinal epithelium. III. Evidence from columnar, enteroendocrine and mucous cells in the adult mouse. Am J Anat 160:77–91
8. Blok PHHM, Manni JJ, Van der Broek P, Van Haelst UJGM, Sloof JL (1985) Carcinoid of the larynx: a report of three cases and a review of the literature. Laryngoscope 95:715–719
9. Bussolati G, Gugliotta P, Sapino A, Eusebi V, Lloyd RV (1985) Chromogranin reactive endocrine cells in argyrophilic carcinomas ("carcinoids") and normal tissue of the breast. Am J Pathol 120:186–192
10. Bussolati G, Papotti M, Sapino A, Gugliotta P, Ghiringhello B, Azzopardi JG (1987) Endocrine markers in argyrophilic carcinomas of the breast. Am J Surg Pathol 11:248–256
11. Chejfec G, Capella C, Solcia E, Jao W, Gould VE (1985) Amphicrine cells, dysplasias, and neoplasias. Cancer 56:2683–2690
12. Clayton F, Ordonez NF, Sibley RK, Hanssen G (1982) Argyrophilic breast carcinomas. Evidence of lactational differentation. Am J Surg Pathol 6:323–333
13. Cohle SD, Tschen JA, Smith FE, Lane M, McGavran MH (1979) ACTH-secreting carcinoma of the breast. Cancer 43:2370–2376
14. Cox WF, Peirce GB (1982) The endocrine origin of the endocrine cells of an adenocarcinoma of the colon of the rat. Cancer 50:1530–1538
15. Cubilla AL, Woodruff JM (1977) Primary carcinoid tumor of the breast. A report of eight patients. Am J Surg Pathol 1:283–292

16. DeLellis RA, Tischler AS, Wolfe HJ (1984) Multidirectional differentiation in neuroendocrine neoplasms. J Histochem Cytochem 32:899–904
17. di Sant'Agnese PA, de Mesy Jensen KL (1984) Somatostatin and/or somatostatin-like immunoreactive endocrine-paracrine cells in the human prostate gland. Arch Pathol Lab Med 108:693–696
18. di Sant'Agnese PA, de Mesy Jensen KL, Churukian CJ, Agarwal MM (1985) Human prostatic endocrine- paracrine (APUD) cells. Distributional analysis with a comparison of serotonin and neuron-specific enolase immunoreactivity and silver stains. Arch Pathol Lab Med 109:607–612
19. Duvall E, Johnston A, McLay K, Piris J (1983) Carcinoid tumour of the larynx. A report of two cases. J Laryngol Otol 97:1073–1080
20. Fetissof F, Braudet P, Arbeille B, Penot J, Marboeuf Y, Le Roux J, Guilloteau D, Beaulieu J-L (1986) Calcitonin-secreting carcinoma of the prostate. An immunohistochemical and ultrastructural analysis. Am J Surg Pathol 10:702–710
21. Feun LG, Savaraj N, Legha SS, Silva EG, Benjamin RS, Burgess MA (1988) Chemotherapy for metastatic Merkel cell carcinoma. Review of the M.D. Anderson Hospital's experience. Cancer 62:683–685
22. Frigerio B, Capella C, Eusebi V, Tenti P, Azzopardi JG (1983) Merkel cell carcinoma of the skin: the structure and origin of normal Merkel cells. Histopathology 7:229–249
23. Galante L, Gudmundsson TV, Matthews EW, Tse A, Williams ED, Woodhouse NJY, MacIntyre I (1968) Thymic and parathyroid origin of calcitonin in man. Lancet 2:537–539
24. Gersall DJ, Mazoujian G, Mutch DG, Rudloff MA (1988) Small cell undifferentiated carcinoma of the cervix. A clinicopathologic, ultrastructural, and immunocytochemical study of 15 cases. Am J Surg Pathol 12:684–698
25. Ghali VS, Garcia RL (1984) Prostatic adenocarcinoma with carcinoidal features producing adrenocorticotropic syndrome. Immunohistochemical study and review of the literaure. Cancer 54:1043–1048
26. Gnepp DR, Ferlito A, Hyams V (1983) Primary anaplastic small cell (oat cell) carcinoma of the larynx. Review of the literature and report of 18 cases. Cancer 51:1731–1745
27. Gould VE, Banner BF, Baerwaldt M (1981) Neuroendocrine neoplasms in unusual primary sites. Diagn Histopathol 4:263–277
28. Gould VE, Lee I, Wiedenmann B, Moll R, Chejfec G, Franke WW (1986) Synaptophysin: a novel marker for neurons, certain neuroendocrine cells and their neoplasms. Hum Pathol 17:979–983
29. Gould VE, Moll R, Moll I, Lee I, Franke WW (1985) Neuroendocrine (Merkel) cells of the skin: hyperplasias, dysplasias and neoplasms. Lab Invest 52:334–353
30. Hales M, Rosenau W, Okerlund MD, Galante M (1982) Carcinoma of the thyroid with a mixed medullary and follicular pattern: Morphologic, immunohistochemical and clinical laboratory studies. Cancer 50:1352–1359
31. Holm R, Sobrinho-Simoes M, Nesland JM, Sambade C, Johannessen JV (1987) Medullary thyroid carcinoma with thyroglobulin immunoreactivity. A special entity? Lab Invest 57:258–268
32. Hosking DH, Bowman DM, McMorris SL, Ramsey EW (1981) Primary carcinoid of the testis with metastases. J Urol 125:255–256
33. Hyman VJ, Batsakis JG, Michaels L (1988) Tumors of the Upper Respiratory Tract. Atlas of Tumor Pathology, Second Series, Fascicle 25. Washington, DC, Armed Forces Institute of Pathology, pp 240–248
34. Inoue M, Ueda G, Yamasaki M, Tanaka Y, Hiramatsu K (1984) Immunohistochemical demonstration of peptide hormones in endometrial carcinomas. Cancer 54:2127–2131
35. Kameya T, Shimosato Y, Adachi I, Abe K, Ebihara S, Ono I (1980) Neuroendocrine carcinoma of the paranasal sinus. A morphological and endocrinological study. Cancer 45:330–339
36. Kaneko H, Hojo H, Ishikawa S, Yamanouchi H, Sumida T, Saito R (1978) Norepinephrine-producing tumors of bilateral breasts. A case report. Cancer 41:2002–2007
37. Matsuyama MJ, Inoue T, Ariyoshi Y, Doi M, Suchi T, Sato T, Tashiro K, Chihana T (1979) Argyrophilic cell carcinoma of the uterine cervix with ectopic production of ACTH, β-MSH, serotonin, histamine and amylase. Cancer 44:1813–1823
38. Micheau C, Guerinot F, Bohuon C, Brugere J (1973) Dopamine β-hydroxylase and catecholamines in an olfactory esthesioneuroblastoma. Cancer 35:1309- 1312
39. Miettinen M (1987) Synaptophysin and neurofilament proteins as markers for neuroendocrine tumors. Arch Pathol Lab Med 111:813–818
40. Morgello S, Schwartz E, Horwith M, King ME, Gorden P, Alonso DR (1988) Ectopic insulin production by a primary ovarian carcinoid. Cancer 61: 800–805
41. Murphy GF, Pilch BZ, Dickersin R, Goodman ML, Nadol JB (1980) Carcinoid tumor of the middle ear. Am J Clin Pathol 73:816–822
42. Nesland JM, Holm R, Johannessen JV, Gould VE (1988) Neuroendocrine differentiation in breast lesions. Pathol Res Pract 183:214–221

43. Nesland JM, Menoli VA, Holm R, Gould VE, Johannessen JV (1985) Breast carcinomas with neuroendocrine differentiation. Ultrastruct Pathol 8:225- 240

44. Ordonez NG, Ayala AG, Sneige N, Mackay B (1982) Immunohistochemical demonstration of multiple neurohormonal polypeptides in a case of pure testicular carcinoid. Am J Clin Pathol 78:860-864

45. Riddle PJ, Font RL, Zimmerman LE (1982) Carcinoid tumour of the eye and orbit. A clinicopathologic study of 15 cases with histochemical and electron microscopic observations. Hum Pathol 13: 459-464

46. Robboy SJ, Scully RE (1980) Strumal carcinoid of the ovary: An analysis of 50 cases of a distinctive tumor composed of thyroid tissue and carcinoid. Cancer 46:2019-2034

47. Rode J, Dhillon AP, Thompson RJ, Craig R, Leathem A (1986) Neuroendocrine cells in thymus and thymic carcinoid. J Pathol 148:105A (abstract)

48. Rodiquez FH Jr, Sorma DP, Lunseth JH (1982) Goblet cell carcinoid of the appendix. Hum Pathol 13:286-288

49. Ro JY, Tetu B, Ayala AG, Ordonez NG (1987) Small cell carcinoma of the prostate. II. Immunohistochemical and electron microscopic studies of 18 cases. Cancer 59:977-982

50. Rosai J, Higa E, Davie J (1972) Mediastinal endocrine neoplasm in patients with multiple endocrine adenomatosis: A previously unrecognized association. Cancer 29:1075-1083

51. Rosai J, Levine GD (1976) Tumors of the Thymus. Atlas of Tumor Pathology, Second Series, Fascicle 13. Washington, DC, Armed Forces Institute of Pathology, pp 167-181

52. Scully RE (1979) Tumors of the Ovary and Maldeveloped Gonads. Atlas of Tumor Pathology, Second Series, Fascicle 16. Washington, DC, Armed Forces Institute of Pathology

53. Scully, RE, Aguirre P, DeLellis RA (1984) Argyrophilia, serotonin and peptide hormones in the female genital tract and its tumors. Int J Gynecol Pathol 3:51-70

54. Sibley RK, Dahl D (1985) Primary neuroendocrine (Merkel cell?) carcinoma of the skin. II. An immunocytochemical study of 21 cases. Am J Surg Pathol 9:109-116

55. Sibley RK, Dehner LP, Rosai J (1984) Primary neuroendocrine (Merkel cell?) carcinoma of the skin. I. A clinicopathologic and ultrastructural study of 43 cases. Am J Surg Pathol 9:95-105

56. Silva EG, Kott MM, Ordonez NG (1984) Endocrine carcinoma intermediate cell type of the uterine cervix. Cancer 54:1705-1713

57. Singh W, Ramage C, Bedst P, Angus B (1980) Nasal neuroblastoma secreting vasopressin. A case report. Cancer 45:961-966

58. Sporrong B, Falkmer S, Robboy SJ, Alumets J, Hakanson R, Ljungberg O, Sundler F (1982) Neurohormonal peptides in ovarian carcinoids. An immunohistochemical study of 81 primary carcinoids of intra-ovarian metastases from six mid-gut carcinoids. Cancer 49:68-74

59. Stagno PA, Petras RE, Hart WR (1987) Strumal carcinoids of the ovary. An immunohistologic and ultrastructural study. Arch Pathol Lab Med 111: 440-446

60. Stahl RE, Sidhu GS (1979) Primary carcinoid of the kidney. Light and electron microscopic study. Cancer 44:1345-1349

61. Sullivan JL, Packer JT, Bryant M (1981) Primary malignant carcinoid of the testis. Arch Pathol Lab Med 105:515-517

62. Taxy JB, Bharani NK, Mills SE, Frierson HF Jr, Gould VE (1986) The spectrum of olfactory neural tumors. A light microscope immunohistochemical and ultrastructural analysis. Am J Surg Pathol 10: 687-695

63. Tetu B, Ro JY, Ayala AG, Johnson DE, Logothetis CJ, Ordonez NG (1987) Small cell carcinoma of the prostate. Part 1: A clinicopathologic study of 20 cases. Cancer 59:1803-1809

64. Ueda G, Sato Y, Yamasaki M, Inoue M, Hiramatsu K, Keiichi K, Amino N, Miyai K (1978) Strumal carcinoid of the ovary. Histological, ultrastructural, immunohistological studies with anti-human thyroglobulin. Gynecol Oncol 6:411-419

65. Vuitch MF, Mendelsohn G (1981) Relationship of ectopic ACTH production to tumor differentiation: A morphologic and immunohistochemical study of prostatic carcinoma with Cushing's syndrome. Cancer 47:296-299

66. Wada A, Ishigur S, Tateishi R, Ishikawa O, Matsui Y (1983) Carcinoid tumor of the gallbladder associated with adenocarcinoma. Cancer 51:1911-1917

67. Warner TFCS (1978) Carcinoid tumours of the uterine cervix. J Clin Pathol 31:990-995

68. Warner TFCS, Uno H, Hafez GR, Burgess J, Bolles C, Lloyd RV, Oka M (1983) Merkel cells and Merkel cell tumors. Ultrastructural immunocytochemistry and review of the literature. Cancer 52: 238-245

68a. Wenig BM, Hyams VJ, Heffner DK (1988) Moderately differentiated neuroendocrine carcinoma of the larynx. A clinicopathologic study of 54 cases. Cancer 62:2658-2676.

69. Wenk RE Bhagavan BS, Levy R, Miller D, Weisburger W (1977) Ectopic ACTH, prostatic oat cell carcinoma and marked hypernatremia. Cancer 40: 296-299

70. Wick MR, Carney JA, Bernatz P, Brown LR (1982) Primary mediastinal carcinoid tumors. Am J Surg Pathol 6:195–205
71. Wick MR, Rosai J (1988) Neduroendocrine neoplasms of the thymus. Pathol Res Pract 183:188–199
72. Wick MR, Scheithauer BW (1984) Thymic carcinoid: A histologic, immunohistochemical and ultrastructural study of 12 cases. Cancer 53:475–484
73. Wick MR, Scott RE, Li C-Y, Carney JA (1980) Carcinoid tumor of the thymus: a clinicopathogic report of seven cases with a review of the literature. Mayo Clin Proc 55:246–254
74. Wilson BS, Lloyd RV (1984) Detection of chromogranin in neuroendocrine cells with a monoclonal antibody. Am J Pathol 115:458–468
75. Woodard BH, Eisenbarth G, Wallace NR, Mossler JA, McCarty KS Jr (1981) Adrenocorticotropin production by a mammary carcinoma. Cancer 47: 1823–1827
76. Woodruff JM, Huvos AG, Erlandson RA, Shah JP, Gerold FP (1985) Neuroendocrine carcinomas of the larynx. A study of two types: one of which mimics thyroid medullary carcinoma. Am J Surg Pathol 9:771–790

CHAPTER 11
Ectopic Hormone Syndromes

Ectopic or inappropriate hormone production (paraneoplastic endocrine syndrome) is the synthesis of hormones by tumors from tissues that do not normally produce the hormone (3,22). This is in contrast to eutopic or entopic hormone production by tumors whose precursor tissues normally produced the hormone. These definitions of ectopic and eutopic hormone production remain valid. However, recent advances in receptor biology, immunohistochemistry, radioimmunoassay, and cell culture techniques have shown that polypeptide hormones are more widely distributed in normal tissues than was previously recognized. Many nonneuroendocrine tissues and derived tumors can also produce polypeptide hormones with multidirectional differentiation (13).

Specific criteria have been established to implicate specific neoplasms as the source of ectopic hormones (3):

1. Clinically overt endocrine syndrome or elevated circulating hormone lines in a patient with a tumor.
2. Cessation of endocrine abnormality after ablation of the tumor.
3. Presence of an arteriovenous gradient in the tumor vascular bed.
4. Hormone concentration gradient between the tumor and adjacent nonneoplastic tissue.
5. Documentation of hormone synthesis by tumor cells in culture and/or demonstration of appropriate mRNA from tumor cells indicating hormone synthesis by the tumor cells.

Specific Syndromes of Ectopic Hormone Production

Some of the major syndromes associated with ectopic hormone production are summarized in Table 11.1. While ectopic production of many hormones can lead to specific clinical syndromes, in many cases ectopic production of polypeptides such as calcitonin, enkephalins, and neurotensin may not be associated with specific clinical syndromes.

Ectopic Hypothalamic Hormone Production

Growth Hormone Releasing Hormone

Growth hormone releasing activity in bronchial endocrine tumors derived from acromegalic patients was first described in 1979, although the suggestion that pulmonary endocrine tumors may cause acromegaly was made two decades earlier (41,56). The growth hormone releasing hormone (GHRH) produced by ectopic sources such as pancreatic tumors is identical to the hypothalamic peptide.

GHRH production as demonstrated by immunohistochemistry in normal tissues has been found only in the central nervous system in the ventromedial nucleus and/or nucleus of the hypothalamus (1,40). More than 30 extracranial tumors producing GHRH ectopically have been reported, including pancreatic, lung, jejunal, retroperitoneal, and adrenal pheochromocytomas (2,40).

Table 11.1. Ectopic hormone production by tumors and syndromes

Hormone	Example of tumor(s)	Clinical syndrome(s), disease(s)
Hypothalamic hormones		
Antidiuretic hormone	Lung carcinoma	Syndrome of inappropriate ADH production
Oxytocin	Lung carcinoma	
Corticotropin releasing hormone	Lung, pancreatic carcinoma	Cushing's
Growth hormone releasing hormone	Pancreas, lung carcinoma, pheochromocytoma	Acromegaly, gigantism
Thyrotropin releasing hormone	Lung carcinoma and trophoblastic disease	Hyperthyroidism (rare)
Anterior pituitary hormones		
ACTH	Lungs islet cells, medullary thyroid carcinoma, pheochromocytoma	Cushing's
Prolactin	Lung carcinoma	Galactorrhea
Growth hormone	Lung carcinoma	Acromegaly
Gonadotropin	Lung carcinoma	
Thyrotropin	Trophoblastic disease	Thyrotoxicosis
Gastroenteropancreatic hormones		
Insulin	Miscellaneous neoplasms	Hypoglycemia
Insulinlike		Hypoglycemia
Gastrin		Hypergastrinemia
Vasoactive intestinal polypeptide		Verner-Morrison
Serotonin		Carcinoid syndrome
Miscellaneous hormones		
Chorionic gonadotropin	Throphoblastic disease	Usually none
Calcitonin	Lung carcinoma, mixed cell carcinoma, pheochromocytosis, carcinoids	
Parathyroid hormone	Lung carcinomas	Hypercalcemia
Parathyroid hormone related peptide	Lung carcinomas	Hypercalcemia
Erythropoietin	Renal carcinomas	Erythrocytosis, hypertension

Intracranial lesions including hypothalamic gangliocytomas have also been associated with GHRH production causing acromegaly (1). Patients with ectopic GHRH production ranged from 15 to 63 years with a mean of 40 years (40). There is a high female predominance (F:M ratio of 22:8). Acromegalic features in patients with extracranial GHRH-producing tumors are indistinguishable from those of classic acromegalics. Associated symptoms due to other hormones, including galactorrhea with or without amenorrhea with high serum prolactin levels, Cushing's and Zollinger-Ellison syndromes, and hyperinsulinemic hypoglycemia, have been reported (40). The pituitary gland often shows growth hormone cell hyperplasia, in many cases with an enlargement of the sella turcica or an intrasellar tumor-like mass.

The light microscopic features of these tumors include solid nests or a tubular pattern of uniform cells with round to oval nuclei and granular cytoplasm. Occasional tumors may consist of spindle cells. Ultrastructural features include dense-core secretory granules ranging from 100 to 250 nm in diameter and well-developed cytoplasmic organelles.

Immunostaining shows GHRH immunoreactivity in varying numbers (Fig. 11.1). Other peptides, including somatostatin, gastrin, calcitonin, gastric releasing peptide, pancreatic polypeptide, glucagon, and insulin, have also been detected (40).

A significant number of patients with ectopic pancreatic GHRH have multiple endocrine neoplasia I syndrome (Chapter 12) and the pancreatic tumors are usually the source of the GHRH (2,40).

Corticotropin Releasing Hormone

Corticotropin releasing hormone (CRH) is a 41-amino-acid peptide. Ectopic CRH production has been reported in lung neuroendocrine carcinomas (9), prostatic carcinomas (5), medullary thyroid carcinomas, and pheochromocytomas (10). Some

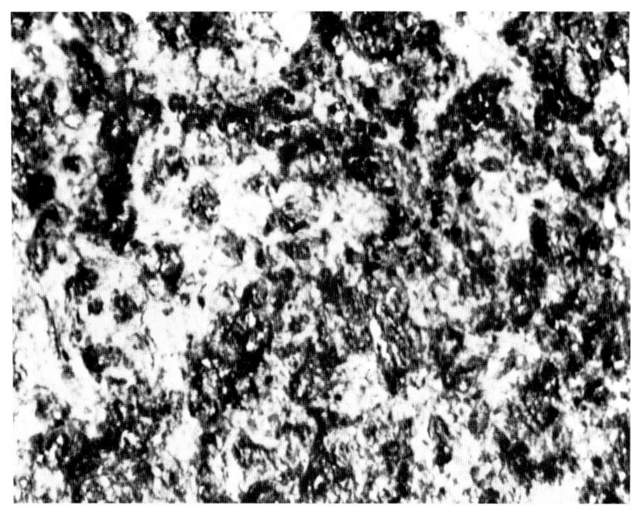

Figure 11.1. Pancreatic growth hormone releasing hormone (GHRH) produced by a tumor of the pancreas metastatic to the liver. The patient also had pituitary hyperplasia and increased serum growth hormone levels. Immunostaining of the liver tumor shows diffuse strongly positive immunoreactivity for GHRH.

cases of ectopic CRH production have been associated with ectopic ACTH production by the same tumor (52).

Other Hypothalamic Releasing Hormones

Thyrotropin releasing hormone has been identified in extracts of various human neoplasms including small-cell carcinomas of lung, melanoma, adenocarcinoma of lung, breast carcinoma, leiomyosarcoma, and carcinoma of the gallbladder (54). These cases were not associated with excess FSH production. Somatostatin production has been reported in cultured cells from small-cell lung carcinoma (47).

Ectopic Production of Pituitary Hormones

Ectopic ACTH

This is one of the most common ectopic hormones produced by various tumors. The most common neoplasms associated with the syndrome include carcinoma of lung (50%) (Fig. 11.2), thymic tumors (10%), pancreatic endocrine tumors (10%), pheochromocytomas, paragangliomas, and related tumors (31). In patients with aggressive neoplasms such as small-cell carcinomas of the lung, which can convert proopiomelanocortin to ACTH, the signs of Cushing's syndrome may be absent or minimal because of the rapid fatal outcome of these tumors. However, slower-growing tumors such as thymic neuroendocrine neoplasms, pheochromocytomas, and medullary thyroid carcinomas may be associated with Cushing's syndrome months to years before the tumor is diagnosed (31). Some tumors producing ACTH may also produce CRH, leading also to pituitary ACTH cell hyperplasia.

Ectopic Vasopressin

Vasopressin or antidiuretic hormone (ADH) is normally produced by the hypothalamus and stored in the posterior pituitary. Inappropriate ADH syndrome associated with ectopic ADH production is commonly associated with small-cell carcinoma of the lung. However, squamous cell and adenocarcinoma of the lung may also lead to the inappropriate ADH syndrome. Rare cases of prostatic, gastrointestinal tract, and adrenal cortical carcinoma have been reported in association with inappropriate ADH production (6,16,31,43). Inappropriate ADH syndrome is associated with hyponatremia, hyperkalemia, renal sodium loss, and inappropriately high urine osmolality. However, in the absence of excess water intake, excess ADH produces no known symptoms (31). Lokich has noted that patients with small-cell carcinomas of the lung and inappropriate ADH secretion often have brain metastases (26).

acteristics included amine production and/or amine precursor uptake, amino acid decarboxylase, high esterase/cholinesterone levels, high α-glycerophosphate dehydrogenase activity, presence of secretory granules, and specific peptides in cells and tumors (33,34). Most of the cells of the DNS are not derived directly from the neural crest, so there is not a common histogenesis; however, many of the tumors producing ectopic hormones are members of the APUD system or DNS and share many of the characteristics of cells and tumors of this system. This might suggest a common mechanism of derepression among the related cells. Many non-neuroendocrine tumors, however, also produce ectopic hormones and this is not adequately explained by the APUD concept.

The genetic theory suggests that derepression or mutations developing during tumorigenesis may lead to ectopic hormone production. This gene may be unmasked or activated or genetic mutations may occur during development of neoplasias (18,44). The genetic mutation theory would suggest that the gene products and polypeptide products of ectopically produced hormones may be abnormal in many cases since mutations would be random events. Experimental evidence for the derepression theory is lacking.

Odell and his co-workers have suggested that most normal tissues produce specific peptides whether they are endocrine or not and that neoplastic development may result in the amplification of these genes or decreased repression. This concept is supported by some experimental evidence such as the production of chorionic gonadotropins by many neoplasms and normal tissues (32). Recent studies on the proopiomelano-cortin (POMC) gene expression in normal tissues and tumors indicate that many tissues also produce POMC (12,25). Normal tissues express POMC mRNA and most pheochromocytomas and a lung squamous cell, adenocarcinoma, and squamous cell laryngeal carcinoma studied by DeBold and colleagues also produced POMC mRNA (12). Interestingly, Cushing's syndrome due to ectopic ACTH secretion was only infrequently expressed in these patients.

Finally, the sponge theory proposed that selective uptake of peptide by tumors would explain ectopic hormones found in these neoplasms (51). However, a large body of evidence from cell culture studies and immunochemical and hybridization analyses have supported direct synthesis of the ectopic peptides and mRNAs by the neoplasms (12,25,49).

References

1. Asa SL, Scheitauer BW, Bilbao JM, Horvath E, Ryan N, Kovacs K, Randall RV, Laws ER Jr, Singer W, Linfoot JA, Thorner MO, Vale W (1984) A case for hypothalamic acromegaly: a clinicopathological study of six patients with hypothalamic gangliocytomas producing growth hormone-releasing factor. J Clin Endocrinol Metab 58:796–803
2. Asa SL, Singer W, Kovacs K, Horvath E, Murray D, Colapinto N, Thorner MO (1987) Pancreatic endocrine tumour producing growth hormone-releasing hormone associated with multiple endocrine neoplasia type I syndrome. Acta Endocrinol 115: 331–337
3. Baylin SB, Mendelsohn G (1980) Ectopic (inappropriate) hormone production by tumors: Mechanisms involved and the biological and clinical implications. Endocrinol Rev 1:45–77
4. Beck C, Burger HG (1972) Evidence for the presence of immunoreactive growth hormone in cancers of the lung and stomach. Cancer 30:75–79
5. Belsky JL, Cuello B, Swanson LW, Simmons DM, Jarrett RM, Braza F (1985) Cushing's syndrome due to ectopic production of corticotropin-releasing factor. J Clin Endocrinol Metab 60:496–500
6. Bondy PK, Gilby ED (1982) Endocrine function in small cell undifferentiated carcinoma of the lung. Cancer 50:2147–2153
7. Braunstein GD, Vaitukaitis JL, Barbone PP (1973) Ectopic production of human chorionic gonadotropin by lung neoplasms. Ann Intern Med 88:805
8. Broadhus AE, Mangin M, Ikeda K, Insogna KL, Weir EC, Burtis WJ, Stewart AF (1988) Humoral hypercalcemia of cancer. Identification of a novel parathyroid hormone-like peptide. N Engl J Med 319:556–563
9. Carey RM, Varma SK, Drake CR, Thorner MO, Kovacs K, Riview J, Vale W (1984) Ectopic secretion of corticotropin-releasing factor as a cause of Cushing's syndrome: a clinical, morphological and biochemical study. N Engl J Med 311:13–20
10. Cunnah D, Jessop DS, Millar JGB, Neville E, Coates P, Doniach I, Hale AC, Besser GM, Rees LH (1986) An adrenal medullary tumour presenting with Cushing's syndrome associated with elevated levels of human corticotropin-releasing factor in plasma. J Endocrinol 108:272

11. Daughaday WH, Emanuele MA, Brooks MH, Barbato AL, Kapadia M, Rotwein P (1988) Synthesis and secretion of insulin-like growth factor II by a leiomyosarcoma with associated hypoglycemia. N Engl J Med 319:1434–1440
12. DeBold CR, Menefee JK, Nicholson WE, Orth DN (1988) Proopiomelanocortin gene is expressed in many normal human tissues and in tumors not associated with ectopic adrenocorticotropin syndrome. Mol Endocrinol 2:862–876
13. DeLellis RA, Tischler AS, Wolfe HJ (1984) Multidirectional differentiation in neuroendocrine neoplasms. J Histochem Cytochem 32:899–904
14. Drezner MK, Feinglos MN (1977) Osteomalacia due to a 1α,25-dihydroxycholecalciferol deficiency. J Clin Invest 60:1046–1053
15. Faiman C, Colwell JA, Ryan RJ, Hershman JM, Shields TW (1967) Gonadotropin secretion from a bronchogenic carcinoma. Demonstration by radioimmunoassay. N Engl J Med 277:1395–1399
16. Falchuk KR (1973) Inappropriate antidiuretic hormone-like syndrome associated with an adrenocortical carcinoma. Am J Med 266:393–395
17. Field JB, Keen H, Johnson P, Herring B (1963) Insulin-like activity of nonpancreatic tumors associated with hypoglycemia. J Clin Endocrinol Metab 23:1229–1236
18. Gellhorn A (1969) Ectopic hormone production in cancer and its implications for basic research on abnormal growth. Adv Intern Med 15:299–316
19. Gorden P, Hendricks CM, Kahn CR, Megyesi K, Roth J (1981) Hyperglycemia associated with non-islet-cell tumor and insulin-like growth factors. N Engl J Med 305:1452–1455
20. Hammond D, Winnick S (1974) Paraneoplastic erythrocytosis and ectopic erythropoietins. Ann NY Acad Sci 230:219–227
21. Heath E, Edis AJ (1979) Pheochromocytoma associated with hypercalcemia and ectopic secretion of calcitonin. Ann Intern Med 91:208–210
22. Imura H (1980) Ectopic hormone syndromes. Clin Endocrinol Metab 9:235–260
23. Johannessen JV, Gould VE (1980) Neuroendocrine skin carcinoma associated with calcitonin production. A Merkel cell carcinoma. Hum Pathol 2:586–589
24. Kahn CR, Rosen SW, Weintraub BD, Fajans SS, Gorden P (1977) Ectopic production of chorionic gonadotropin and its subunits by islet cell tumors. A specific marker for malignancy. N Engl J Med 297:565–569
25. Lacaza-Masmonteil T, DeKeyzer Y, Luton J-P, Kahn A, Bertagna X (1987) Characterization of proopiomelanocortin transcripts in human nonpituitary tissues. Proc Natl Acad Sci USA 84:7261–7265
26. Lokich JJ (1982) The frequency and clinical biology of the ectopic hormone syndromes of small cell carcinoma. Cancer 50:2111–2114
27. Mangin M, Webb AC, Dreyer B, Posillico JT, Ikeda K, Weir EC, Stewart AF, Bander NH, Milstone L, Barton DE, Francke U, Broadhus AE (1988) Identification of a cDNA extracting a parathyroid hormone-like peptide from a human tumor associated with humoral hypercalcemia of malignancy. Proc Natl Acad Sci USA 85:597–601
28. McFadzean AJS, Yeung RTT (1969) Further observations on hypoglycemia in hepatocellular carcinoma. Am J Med 47:220–235
29. Morley JE, Jacobson RJ, Melamed J, Hershman JM (1976) Choriocarcinoma as a cause of thyrotoxicosis. Am J Med 60:1036–1040
30. Mundy GR, Ibbotson KJ, D'Souza SM, Simpson EL, Jacobs JW, Martin TJ (1984) The hypercalcemia of cancer: clinical implications and pathogenic mechanisms. N Engl J Med 310:1718–1727
31. Odell WD (1985) Hormonal manifestation of cancer. In: Wilson JD, Foster DW (eds) Williams Textbook of Endocrinology, 7th ed. Philadelphia, Saunders, pp 1327–1344
32. Odell WD, Wolfsen A, Yoshimoto Y (1977) Ectopic peptide synthesis: A universal concomitant of neoplasia. Trans Assoc Am Phys 90:204
33. Pearse AGE (1968) Common cytochemical and ultrastructural characteristics of cells producing polypeptide hormones (the APUD series) and their relevance to thyroid and ultimobranchial C cells and calcitonin. Proc Soc London Ser B 170:71–80
34. Pearse AGE (1986) The diffuse neuroendocrine system: Peptides, amines, placodes and the APUD theory. Prog Brain Res 68:25–31
35. Plimpton CH, Gellhorn A (1956) Hypercalcemia in malignant disease without evidence of bone destruction. Am J Med 21:750–759
36. Rees LH, Ratcliffe JG (1974) Ectopic hormone production by nonendocrine tumours. Clin Endocrinol 3:263–299
37. Rosen SW, Weintraub BD (1974) Ectopic production of the isolated alpha subunit of the glycoprotein hormones. A quantitative marker in certain cases of cancer. N Engl J Med 290:1441–1447
38. Rosen SW, Weintraub BD, Vaitukaitis JL, Sussman HH, Hershman JM, Muggia FM (1975) Placental proteins and their subunits as tumor markers. Ann Intern Med 82:71–83
39. Said SI, Faloona GR (1975) Elevated plasma and tissue levels of vasoactive intestinal polypeptide in the watery diarrhea syndrome due to pancreatic, bron-

chogenic and other tumors. N Engl J Med 293: 155–160
40. Sano T, Asa SL, Kovacs K (1988) Growth hormone-releasing hormone-producing tumors: clinical, biochemical and morphological manifestations. Endocrinol Rev 9:357–373
41. Schalet SM, Beardwell CG, MacFarlane IA, Ellison ML, Norman CM, Rees LH, Hughes M (1979) Acromegaly due to production of a growth hormone releasing factor by a bronchial carcinoid tumour. Clin Endocrinol 10:61–67
42. Schwartz KE, Wolfsen AR, Forster B (1979) Calcitonin in nonthyroidal cancer. J Clin Endocrinol Metab 49:438–444
43. Sellwood RA, Spencer J, Azzopardi JG, Wapnick S, Welbourn RB, Kulatilake AE (1969) Inappropriate secretion of antidiurectic hormone by carcinoma of the prostate. Br J Surg 56:933–935
44. Shields R (1977) Gene derepression in tumours. Nature 269:752–753
45. Silva OL, Becker KL, Primack A, Doppman JL, Snider RH (1976) Increased serum calcitonin levels in bronchogenic carcinoma. Chest 69:495–499
46. Suva LJ, Winslow GA, Wettenhall REH, Hammonds RG, Moseley JM, Diefenbach-Jagger H, Rodda CP, Kemp BE, Rodriguez H, Chen EY, Hudson PJ, Martin TJ, Wood WI (1987) A parathyroid hormone-related protein implicated in malignant hypercalcemia: cloning and expression. Science 237:893–896
47. Szabo M, Berelowitz M, Pettengill OS, Sorenson GD, Frohman LA (1980) Ectopic production of somatostatin-like immuno- and bioactivity by cultured human pulmonary small cell carcinoma. J Clin Endocrinol Metab 51:978–987
48. Talstad I, Folling I, Boye NP (1974) Hypoglycemia caused by an intrathoracic tumour. Acta Med Scand 196:347–351
49. Tashjian AH Jr, Weintraub BD, Barousky NJ (1973) Subunits of human chorionic gonadotropin: Unbalanced synthesis and secretion by clonal cell strains derived from a bronchogenic carcinoma. Proc Natl Acad Sci USA 70:1419
50. Trump DL, Livingston JN, Baylin SB (1977) Watery diarrhea syndrome in an adult with ganglioneuroma-pheochromocytoma: Identification of vasoactive intestinal peptide, calcitonin, and catecholamines and assessment of their biologic activity. Cancer 40:1526–1532
51. Unger RH, Lockner JV, Eisentraut AM (1964) Identification of insulin and glucagon in a bronchogenic metastasis. J Clin Endocrinol Metab 24:823–831
52. Upton GV, Amatruda TT (1971) Evidence for the presence of tumor peptides with corticotropin-releasing factor-like activity in the ectopic ACTH syndrome. N Engl J Med 285:419–424
53. Weintraub BD, Rosen SW (1973) Ectopic production of the isolated beta subunit of human chorionic gonadotropin. J Clin Invest 52:3135–3142
54. Wilber JF, Spinella P (1984) Identification of immunoreactive thyrotropin-releasing hormone in human neoplasia. J Clin Endocrinol Metab 59: 432–435
55. Yesner R (1978) Spectrum of lung cancer and ectopic hormones. Pathol Annu 13(Part 1):217–240
56. Zafar MS, Mellinger RC, Fine G, Szabo M, Frohman LA (1975) Acromegaly associated with a bronchial carcinoid tumor: Evidence for ectopic production of growth hormone-releasing activity. J Clin Endocrinol Metab 48:66–71

CHAPTER 12

Polyendocrine Disorders

The polyendocrine disorders include diseases generally associated with hypofunction of various endocrine tissues, such as the autoimmune polyglandular syndromes, and conditions associated with hyperplasias and neoplasias with hyperfunctioning endocrine tissues, such as the various multiple endocrine neoplasia syndromes.

Autoimmune Polyglandular Syndromes

Type I or Polyglandular Deficiency Associated with Mucocutaneous Candidiasis

This uncommon disorder is associated with mucocutaneous candidiasis that appears in early childhood (Table 12.1). Subsequent development of idiopathic adrenal insufficiency, hypoparathyroidism, or both conditions can occur. Patients may develop chronic active hepatitis, malabsorption, primary hypoparathyroidism, alopecia, pernicious anemia, and occasionally hypothyroidism. Antibodies against the adrenal glands, parathyroid glands, and gastric parietal cells may be present (20,74). Cases of insulin-dependent diabetes have also been reported. The inheritance may be autosomal recessive and the syndrome does not appear to be HLA associated.

Type II Polyglandular Syndrome

The association of idiopathic Addison's disease, adrenal cortical hypofunction, and lymphocytic thyroiditis is designated as Schmidt's syndrome. The type II syndrome is inherited as an autosomal dominant trait and is associated with insulin-dependent diabetes mellitus in close to half of all patients (20). Myasthenia gravis, celiac disease, pituitary hypogonadism, vitiligo, pernicious anemia, and alopecia may also occur with increased frequency (Table 12.1). The important factors are thought to be an HLA-associated genetic predisposition coupled with environmental effects (21). Autoantibodies to endocrine tissues and to gastric parietal cells may develop. The disease is more common in females and usually involves adults from 20 to 60 years of age (20). The interval between the onset of one endocrinopathy and the diagnosis of the next disease may be very long, extending up to 20 years. The development of overt insulin-dependent diabetes mellitus is usually preceded by autoantibodies by several years (7).

Table 12.1. Autoimmune polyglandular syndromes

	Type I[a]	Type II[a]
Hypoadrenalism	++	++
Hypothyroidism	++	+
Hypoparathyroidism	+	++
Insulin-dependent diabetes mellitus	+	++
Autoantibodies	++	++
HLA-DR associated	−	++
Mucocutaneous candidiasis	−	+
Vitiligo	+	+
Alopecia	+	+
Myasthenia gravis	+	−

[a]++, Common; +, less common; +/−, rare; −, absent

Table 12.2. Multiple endocrine neoplasia.

	Type[a]		
	1	2a (II)	2b (III)
Hyperparathyroidism	++	+	–
Pituitary adenomas	+	–	–
Pancreatic hyperplasia/ adenoma	+	–	–
Thyroid C-cell hyperplasia/ carcinoma	–	++	++
Pheochromocytoma/ medullary hyperplasia	–	++	++
Mucosal neuromas	–	–	+
Lipomas	+/–	–	–
Carcinoid tumors	+/–	–	–
Adrenal cortical and thyroid adenomas	+/–	–	–

[a] ++, Common; +, less common; +/–, rare; –, absent.

Multiple Endocrine Neoplasia (MEN)

The multiple endocrine neoplasia syndromes are inherited as autosomal dominant disorders with a high degree of penetrance (Table 12.2). Sporadic cases occur by mutations frequently with MEN 2b (III). Affected individuals often develop hyperplasia of many endocrine glands, which often precedes neoplastic development. Multifocal hyperplasia or neoplasia involving the adrenal medulla and thyroid is common in patients with MEN and helps to separate sporadic from familial disorders.

Multiple Endocrine Neoplasia 1 (Wermer's Syndrome)

This condition was first described in 1954 (70). The principal features include hyperparathyroidism, pancreatic hyperplasia, and/or adenomas and pituitary adenomas (41). Carcinoid tumors and lipomas occur in some patients. Hyperparathyroidism occurs in most patients with MEN 1 (16,29). Patients with MEN I syndrome usually become symptomatic during the third to fifth decade (37). Morbidity and mortality with MEN 1 is principally related to pancreatic disease, especially gastrin-secreting tumors.

Parathyroid Abnormalities

Hyperparathyroidism occurs in most patients with MEN 1 syndrome. Patients with primary hyperparathyroidism associated with MEN 1 are usually 20 to 30 years younger than patients with sporadic hyperparathyroidism. Most patients with MEN 1-related hyperparathyroidism have parathyroid hyperplasia, although some may have adenomas. Histologically, the parathyroid glands are multinodular with a decrease in stromal lipid content (Fig. 12.1). A few patients may develop parathyroid adenomas. Parathyroid carcinomas are uncommon in these patients.

Patients with MEN 1 who are treated with subtotal parathyroidectomy may have recurrent hypercalcemia and hyperparathyroidism (52).

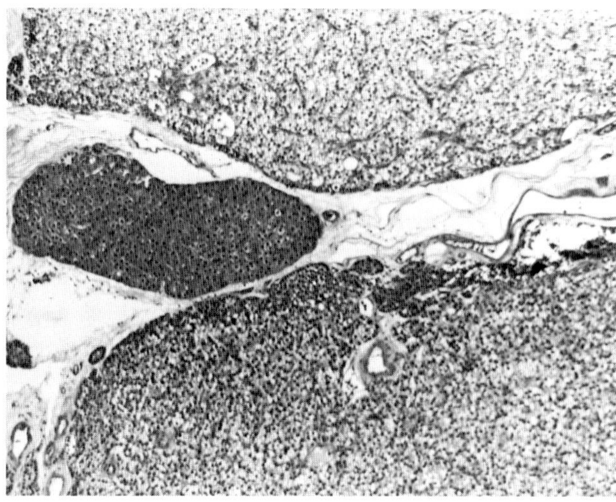

Figure 12.1. Nodular parathyroid hyperplasia in a patient with multiple endocrine neoplasia type 1. Prominent clear cells and oncocytic nodules are present.

Pancreatic Abnormalities

Pancreatic lesions occurring in patients with MEN 1 syndrome include islet cell hyperplasia with nesidioblastosis and pancreatic endocrine neoplasms (47, 65,67). Many of the neoplasms produce gastrin, insulin, or other hormones, while some neoplasms are nonfunctional. Production of ectopic hormones including growth hormone releasing hormone, corticotropin releasing hormone, ACTH, and calcitonin may also occur (4,65).

Gastrinomas are the most common pancreatic endocrine neoplasms in patients with MEN 1. They occur in approximately two-thirds of patients with MEN 1. Zollinger-Ellison syndrome (peptic ulceration associated with gastrinomas) is the major cause of morbidity and mortality in patients with MEN 1 syndrome (5,8,35). The prevalence of MEN 1 in patients with the Zollinger-Ellison syndrome varies from 20 to 60% (11). Insulin-producing tumors are the second most common pancreatic endocrine neoplasms in patients with MEN 1. Some patients may have both gastrin- and insulin-producing tumors with clinical features of both (51). The incidence of malignant insulinomas in patients with MEN 1 has been reported to be greater than in those with sporadic insulinomas (63).

Other pancreatic endocrine tumors commonly developing in patients with MEN 1 include glucagonomas (17) and pancreatic polypeptide-producing tumors (25). Elevated serum levels of pancreatic polypeptide have been noted in patients with MEN 1 (25,34). Elevated pancreatic polypeptide may also be seen in islet cell hyperplasia or in hypergastremia or hyperinsulinemia (37).

The pancreas in patients with MEN 1 may contain multiple tumor nodules that can be seen on gross inspection. Histologic examination of the pancreas shows a spectrum of changes. Multiple nodules ranging from a few millimeters to a few centimeters in diameter may be present as diffuse microadenomatosis. Islet cell hyperplasia with enlarged islets and nesidioblastosis with islet cell proliferation from ducts and small clusters of islet cells admixed with acinar cells may be present (Fig. 12.2). Ultrastructural studies often show the characteristic granule morphology in areas of hyperplasia or nesidioblastosis. The secretory granules in adenomas and carcinomas often do not retain the ultrastructural features of the granules seen in the normal pancreas.

Immunohistochemical staining with broad-spectrum endocrine markers and with specific antibodies against pancreatic endocrine hormones helps to characterize the mixed nature of the proliferating cells in hyperplasia/nesidioblastosis. Adenomas and carcinomas often contain multiple hormones similar to the sporadic tumors. In most cases symptomatic patients may have evidence of excessive production of only one hormone. The most common hormone in MEN 1 pancreatic tumors is pancreatic polypeptide, followed by glucagon and then insulin (31).

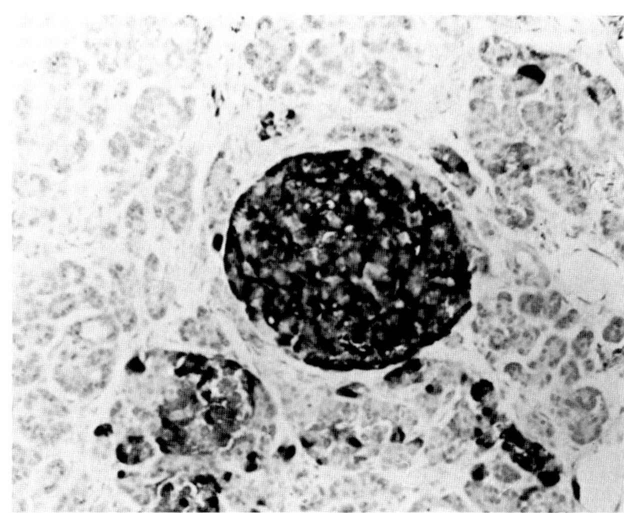

Figure 12.2. Pancreatic tissue from a patient with multiple endocrine neoplasia type 1 stained for chromogranin A. An enlarged islet and small aggregates of islet cells are present.

Pituitary Adenomas

The incidence of pituitary adenomas in patients with MEN 1 is at least 50 to 60% and may be even higher (37,41). Prolactinomas are the most common tumors associated with MEN 1 as well as the most common sporadic pituitary adenomas (38,53, 68). Growth hormone-producing tumors have also been reported in patients with MEN 1 (5). Hyperplasia of GH-producing cells has been observed in association with MEN 1, but this may be related to ectopic production of GHRH by pancreatic or other neoplasms (3,4,57).

The histological and ultrastructural features of the adenomas in patients with MEN 1 are similar to those of pituitary adenomas occurring sporadically. Densely and sparsely granulated GH adenomas and sparsely granulated prolactinomas have been observed. Cases of GH cell hyperplasia have been reported with MEN 1 syndrome. However, many of these cases probably do not represent a genetically related hyperplasia as seen in the nodular hyperplasia associated with familial medullary thyroid carcinoma and pheochromocytomas. These cases probably represent marked pituitary hyperplasia caused by GHRH secretion from a pancreatic endocrine tumor.

Other Endocrine Tumors Associated with MEN 1

Approximately 25 to 40% of patients with MEN 1 in the early studies were reported to have hyperplasia or adenomas of the adrenal cortex (5). The hyperplasia may be secondary to ectopic ACTH production and stimulation of the adrenal cortex. Since adrenal cortical nodules are a common finding at autopsy, the true incidence of adenomas related to MEN 1 is not known.

Carcinoid tumors of the bronchus, duodenum, and thymus have been reported to occur more commonly in MEN 1 patients (42,54,72).

Multiple Endocrine Neoplasia Type 2a (Sipple's Syndrome)

This syndrome is associated with medullary thyroid carcinoma, pheochromocytoma, and hyperparathyroidism (60,64,71). Most patients with MEN 2a develop C-cell hyperplasia and medullary thyroid carcinomas. While adrenal medullary tumors and parathyroid adenomas occur in approximately 50% of cases, patients usually develop C-cell disease before they develop pheochromocytomas. The morbidity and mortality in this syndrome are associated with medullary thyroid carcinoma and pheochromocytomas.

Table 12.3. Sporadic and familial medullary thyroid carcinoma

	Sporadic	Familial
Percent of total cases	80–90%	10–20-%
Associated conditions	None	MEN 2a,2b
C-cell hyperplasia	Rare	Common
Average age of onset	Fourth decade	First decade
Location in thyroid	Unilateral Middle to upper lobe	Bilateral Middle to upper lobe

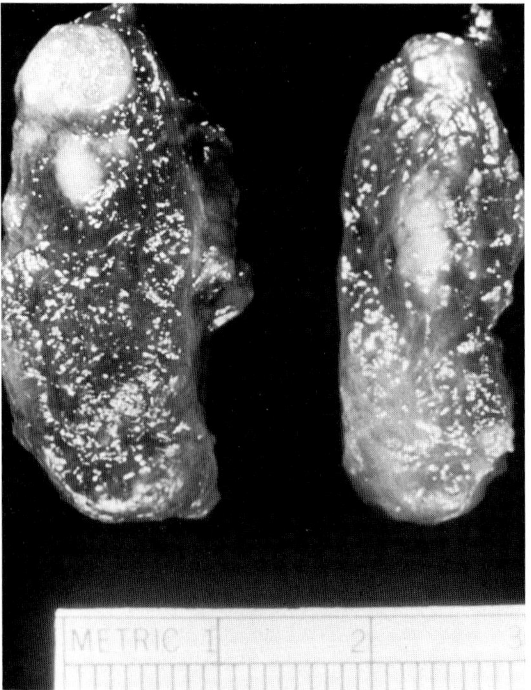

Figure 12.3. Familial medullary thyroid carcinoma. Tumors are present in the upper portion of both left and right thyroid lobes.

Medullary Thyroid Carcinoma

The calcitonin-producing or C cells are present predominantly in the upper half of each thyroid lobe. Although previously considered to be parafollicular in location, ultrastructural studies have revealed that they are intrafollicular (18). Most cases of medullary thyroid carcinoma are sporadic, but about 10 to 20% are familial and are associated with MEN 2a and 2b (Table 12.3). The thyroid disease in these patients usually involves both thyroid lobes (Fig. 12.3). These tumors are preceded by and associated with C-cell hyperplasia.

The progression from normal to hyperplastsic C cells has been elegantly described by DeLellis and Wolfe (18). Histologically, the intrafollicular cells undergo diffuse and nodular hyperplasia. Invasion occurs when the tumor cells break through the basement membrane of the follicles. Most cases of familial medullary thyroid carcinomas have adjacent foci of C-cell hyperplasia (18).

Medullary thyroid carcinomas commonly metastasize to cervical lymph nodes. The extent of metastases is related to tumor size (9). Distant metastases to the liver, mediastinum, lungs, adrenal, and bone occur in advanced stages of the disease. Immunochemical staining for calcitonin (Fig. 12.4) may help to predict the biological behavior of C-cell tumors in specific patients (39,55), since aggressive tumors show focal immunoreactivity for calcitonin

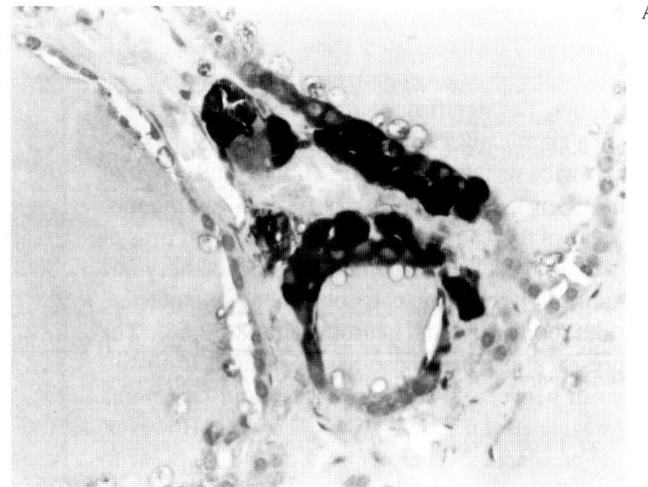

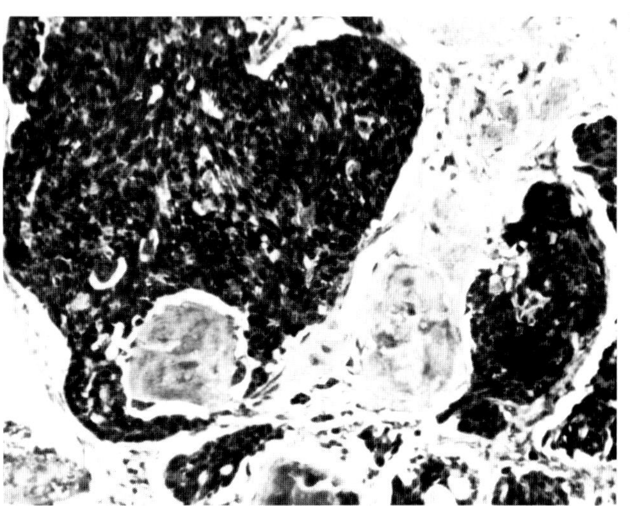

Figure 12.4. (A) C-cell hyperplasia revealed by immunostaining for calcitonin in a patient with MEN 2b. (B) Intense immunoreactivity for calcitonin present in this medullary thyroid carcinoma.

the human calcitonin (C)-cell: A review. Pathol Annu 16(Part 2):25–52
19. DeLellis RA, Wolfe HJ, Gagel RF, Feldman ZT, Miller HH, Gang DL, Reichlin S (1976) Adrenal medullary hyperplasia. A morphometric analysis in patients with familial medullary thyroid carcinoma. Am J Pathol 83:177–196
20. Eisenbarth GS (1985) The immunoendocrinopathy syndromes. In: Wilson JD, Foster DL (eds) Williams Textbook of Endocrinology. Philadelphia, Saunders, pp 1290–1300
21. Eisenbarth GS, Jackson RA (1981) Immunogenetics of polyglandular failure and related diseases in HLA. In: Farid N (ed) Endocrine and Metabolic Disorders. New York, Academic Press, pp 235–264
22. Fontaine J, Le Douarin NM (1977) Analysis of endoderm formation in the avian blastoderm by the use of the quail-chick chimaeras. J Embryol Exp Morphol 41:209–222
23. Freier DT, Thompson NW, Sisson JC, Nishiyama RH, Freitas JE (1977) Dilemmas in the early diagnosis and treatment of multiple endocrine adenomatosis type II. Surgery 82:407–413
24. Freitas JE, Sisson JC, Freier DT, Thompson NW (1978) MEN IIa syndrome: dilemmas in modern management. Semin Nucl Med 8:73–78
25. Friesen SR, Kimmel JR, Tomita T (1980) Pancreatic polypeptide as a screening marker for pancreatic polypeptide adenomas in multiple endocrinopathies. Am J Surg 139:61–72
26. Gorlin RJ, Sedano HO, Vickers RA, Cervenka J (1968) Multiple mucosal neuromas, pheochromocytoma and medullary carcinoma of the thyroid: a syndrome. Cancer 22:293–299
27. Hansen OP, Hansen M, Hansen HH, Rose B (1976) Multiple endocrine adenomatosis of mixed type. Acta Med Scand 200:327–331
28. Hull MT, Warfel KA, Muller J, Higgins JT (1979) Familial islet cell tumors in von Hippel-Lindau's disease. Cancer 44:1523–1526
29. Jung RT, Grant AM, Davie M, Jenkins D (1978) Multiple endocrine adenomatosis (type I) and familial hyperparathyroidism. Postgrad Med J 54: 92–94
30. Keiser HR, Bearen MA, Doppman J, Wells S, Buja LM (1973) Sipple's syndrome: medullary thyroid carcinoma, pheochromocytoma and parathyroid disease: Studies in a large family. Ann Intern Med 78:561–579
31. Kloppel G, Willemer S, Stamm B, Hacki WH, Heitz PU (1986) Pancreatic lesions and hormonal profile of pancreatic tumors in multiple endocrine neoplasia type I. An immunocytochemical study of nine patients. Cancer 57:1824–1832
32. Knudson AG (1985) Hereditary cancer, oncogenes and anti-oncogenes. Cancer Res 45:1437–1443
33. Knudson AG Jr (1975) Genetics of human cancer. Genetics 79(Suppl):305–316
34. Lamers BHW, Diemel M (1982) Basal and postatropine serum pancreatic polypeptide concentrations in familial multiple endocrine neoplasia type I. J Clin Endocrinol Metab 55:774–778
35. Lamers CB, Stadil F, Van Tongeren JH (1978) Prevalence of endocrine abnormalities in patients with the Zollinger-Ellison syndrome and in their families. Am J Med 64:607–612
36. Larsson C, Skogseid B, Oberg K, Nakamura Y, Nordenskjold M (1988) Multiple endocrine neoplasia type 1 gene maps to chromosome 11 and is lost in insulinoma. Nature 332:85–87
37. Leshin M (1985) Multiple endocrine neoplasia. In: Wilson JD, Foster DW (eds) Williams Textbook of Endocrinology. Philadelphia, Saunders, pp 1274–1289
38. Levine JH, Sagel J, Rosebrock G, Gonzalez JJ, Nair R, Rawe S, Powers JM (1979) Prolactin-secreting adenoma as part of the multiple endocrine neoplasia type I (MEN I) syndrome. Cancer 43:2492–2496
39. Lippman SM, Mendelsohn G, Trump DL, Wells SA, Baylin SB (1982) The prognostic and biological significance of cellular heterogeneity in medullary thyroid carcinoma: A study of calcitonin, L-dopa decarboxylase, and histaminase. J Clin Endocrinol Metab 54:233–240
40. Livolsi VA, Feind CR (1979) Incidental medullary thyroid carcinoma in sporadic hyperparathyroidism. An expansion of the concept of C-cell hyperplasia. Am J Clin Pathol 71:595–599
41. Majewski JT, Wilson SD (1979) The MEA-l syndrome: an all or none phenomenon? Surgery 86: 475–484
42. Manes JL, Taylor HB (1973) Thymic carcinoid in familial multiple endocrine adenomatosis. Arch Pathol 95:252–255
43. Mathew CGP, Smith BA, Thorpe K, Wong Z, Royle NJ, Jeffreys AJ, Ponder BAJ (1987) Deletion of genes on chromosome 1 in endocrine neoplasia. Nature 328:524–526
44. Mendelsohn G, Eggleston JC, Weisburger WR, Gann DS, Baylin SB (1978) Calcitonin and histaminase in C-cell hyperplasia and medullary thyroid carcinoma. A light microscopic and immunohistochemical study. Am J Pathol 92:35–52
45. Moyes CD, Alexander FW (1977) Mucosal neuroma syndrome presenting in a neonate. Dev Med Child Neurol 19:518–534
46. Norton JA, Froome LC, Farrell RE, Wells SA Jr (1979) Multiple endocrine neoplasia type IIb. The most aggressive form of medullary thyroid carcinoma. Surg Clin North Am 59:109–118
47. Oliver MH, Drury PL, Van't Hoff W (1983) A case of multiple endocrine adenomatosis (type I) with

nesidioblastosis, termination with an exocrine pancreatic carcinoma. Clin Endocrinol 18:495–503
48. O'Toole K, Fenoglio-Preiser C, Puchparaj N (1985) Endocrine changes associated with the human aging process. III. Effect of age in the number of calcitonin immunoreactive cells in the thyroid gland. Hum Pathol 16:991–1000
49. Pearse AGE (1968) Common cytochemical and ultrastructural characteristics of cells producing polypeptide hormones (the APUD series) and their relevance to thyroid and ultimobrachial C-cells and calcitonin. Proc Soc London Sev B 170:71–80
50. Pearse AGE (1986) The diffuse neuroendocrine system: peptides, amines, placodes and the APUD theory. Prog Brain Res 68:25–31
51. Peurifoy JT, Gomez LG, Thompson JC (1979) Separate pancreatic gastrin cell and beta-cell adenomas: Report of a patient with multiple endocrine adenomatosis type I. Arch Surg 114:956–958
52. Prinz RA, Gamvros OI, Sellu D (1981) Subtotal parathyroidectomy for primary chief cell hyperplasia of the multiple endocrine neoplasia type I syndrome. Ann Surg 193:26–29
53. Prosser PR, Karam JH, Townsend JJ, Forsham PH (1979) Prolactin-secreting pituitary adenomas in multiple endocrine adenomatosis, type I. Ann Intern Med 91:41–44
54. Rosai J, Higa E, Davie J (1972) Mediastinal endocrine neoplasms in patients with multiple endocrine adenomatosis: a previously unrecognized association. Cancer 29:1074–1083
55. Saad MF, Ordonez NG, Guido JJ, Samaan NA (1984) The prognostic value of calcitonin immunostaining in medullary carcinoma of the thyroid. J Clin Endocrinol Metab 59:850–856
56. Saad MF, Ordonez NG, Rashid RK, Guido JJ, Hill CS, Hickey RC, Saaman NA (1984) Medullary carcinoma of the thyroid gland: A study of the clinical features and prognostic factors in 161 patients. Medicine 63:319–342
57. Sano T, Asa SL, Kovacs K (1988) Growth hormone-releasing hormone producing tumors: clinical, biochemical and morphological manifestations. Endocrine Rev 9:357–373
58. Schroder S, Bocker W, Baisch H, Burk CG, Arps H, Meiners I, Kastendieck H, Heitz PU, Kloppel G (1988) Prognostic factors in medullary thyroid carcinomas. Survival in relation to age, sex, stage, histology, immunocytochemistry, and DNA content. Cancer 61:806–816
59. Sikri KL, Varndell IM, Hamid QA, Wilson BS, Kameya T, Ponder BAJ, Lloyd RV, Bloom SR, Polak JM (1985) Medullary carcinoma of the thyroid: An immunocytochemical and histochemical study of 25 cases using eight separate markers. Cancer 56: 2481–2491
60. Sipple JH (1961) The association of pheochromocytoma with carcinoma of the thyroid gland. Am J Med 31:163–166
61. Spandidos DA, Anderson MLM (1989) Oncogenes and onco-suppressor genes: their involvement in cancer. J Pathol 157:1–10
62. Stanbridge EJ (1986) A case for human tumor-suppressor genes. Bioessays 3:252–255
63. Stafanini P, Carboni M, Patrassi N, Basoli A (1974) Beta-islet cell tumors of the pancreas: results of a study of 1,067 cases. Surgery 75:597–609
64. Steiner AL, Goodman AD, Powers SR (1968) Study of a kindred with pheochromocytoma, medullary thyroid carcinoma, hyperparathyroidism, and Cushing's disease: multiple endocrine neoplasia, type 2. Medicine 47:371–409
65. Thompson NW, Lloyd RV, Nishiyama RH, Vinik AI, Strodel WE, Allow MD, Eckhauser FE, Talpos G, Mervak T (1984) MEN I pancreas: a histological and immunohistochemical study. World J Surg 8: 561–574
66. Uribe M, Fenoglio-Preiser CM, Grimes M, Feind C (1985) Medullary carcinoma of the thyroid gland: Clinical, pathological and immunohistochemical features with review of the literature. Am J Surg Pathol 96:577–594
67. Vance JE, Stoll TW, Kitabchi AE, Williams RH, Wood FC Jr (1969) Nesidioblastosis in familial endocrine adenomatosis. JAMA 207:1679–1682
68. Veldhuis JD, Green JE III, Kovacs E, Worgul TJ, Murray FT, Hammond JM (1979) Prolactin-secreting pituitary adenomas: association with multiple endocrine neoplasia, type I. Am J Med 67:830–837
69. Webb TA, Sheps SG, Carney JA (1980) Differences between sporadic pheochromocytoma and phenochromocytoma in multiple endocrine neoplasia, type 2. Am J Surg Pathol 4:121–126
70. Wermer P (1954) Genetic aspects of adenomatosis of endocrine glands. Am J Med 16:363–371
71. Williams ED (1965) A review of 17 cases of carcinoma of the thyroid and phaeochromocytoma. J Clin Pathol 18:288–292
72. Williams ED, Celestin LR (1962) The association of bronchial carcinoid and pluriglandular adenomatosis. Thorax 17:120–127
73. Williams ED, Pollack DJ (1966) Multiple mucosal neuromata with endocrine tumours: a syndrome allied to von Recklinghausen's disease. J Pathol Bacteriol 91:71–80
74. Wirfalt A (1981) Genetic heterogeneity in autoimmune polyglandular failure. Acta Med Scand 210: 7–13

APPENDIX I
Immunochemistry

Immunochemical detection of antigens in tissue sections was first described by Coons and his colleagues, who localized pneumococcal antigens with fluorescent labeled antibodies (9,10). Major advances in this field resulted from the studies of Nakane (53) and Sternberger (73,74). Nakane introduced the enzyme-labeled antibody method and was able to localize pituitary hormones at the light and electron microscopic levels (53). Sternberger used the unlabeled antibody-enzyme bridge and the peroxidase-antiperoxidase complex procedures with diaminobenzidine to improve the sensitivity of these techniques beyond that of the fluorescent methods in his early study of spirochetes (73,74).

The use of immunohistochemical methods in the study of normal and neoplastic endocrine tissues increased greatly in the late 1970s, with many reported studies of the thyroid (14,19), pituitary (40), and general neuroendocrine system (11,59, 68). With the development of monoclonal antibodies (38), many new antibodies were generated for the study and characterization of endocrine tissues (21,48). The numerous books published in recent years on immunochemistry with specific applications to endocrine pathology attest to the importance of this technique in diagnostic pathology (12,13,58,60,72,76).

Immunohistochemical techniques have been widely used to study normal and pathologic endocrine tissue. Immunohistochemical analysis adds a functional dimension to morphological studies. The use of peroxidase and alkaline phosphatase enzymes with various chromogens that can be readily visualized with the light microscope has contributed greatly to the rapid progress in immunocytochemistry. Various methods of immunochemical staining are available. These can be applied for light microscopic and also electron microscopic immunocytochemical analyses.

Immunocytochemical Methods

Direct Methods

This is the earliest immunocytochemical method that was developed (9). With this method the primary antibody is labeled directly with an enzyme system or a fluorescent tag and applied to the section to identify specific antigens (Fig. I.1). Enzyme systems that can be used include horseradish peroxidase, alkaline phosphatase, glucose oxidase, and β-galactosidase. Electron dense labels such as ferritin can be used for ultrastructural studies. The main advantages of the direct procedure are speed and ease of performance. Highly purified antibodies are needed for assurance of specificity.

Indirect or Sandwich Method

Because this method is more sensitive than the direct method, lower concentrations of the primary antibody can be used to detect specific antigens. The application of an unlabeled primary antibody followed by a secondary antibody conjugated to an enzyme system increases the versatility of the assay.

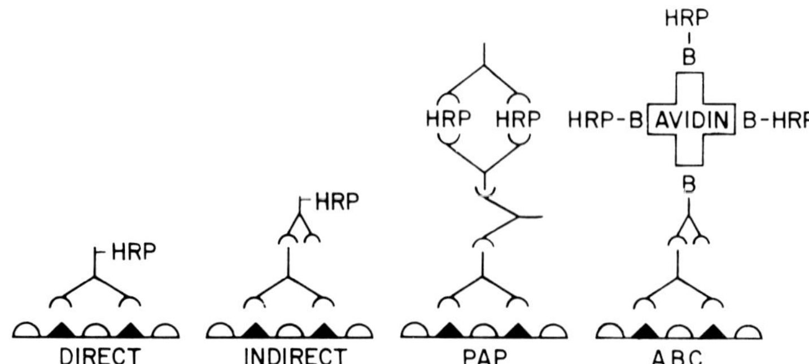

Figure I.1. Methodologic approaches to immunochemical staining. In the direct method the antibody is linked to the detection system enzyme such as horseradish peroxidase. In the indirect method an unlabeled antibody is linked to a secondary antibody conjugated to an enzyme system for increased versatility. The peroxidase-antiperoxidase (PAP) method utilizes a stable immune complex of two antibody molecules and three horseradish peroxidase molecules. The avidin-biotin complex peroxidase (ABC) method takes advantages of the high affinity of avidin for biotin and is probably more sensitive than the PAP method. The antigens in tissues are represented by dark triangles and open semicircles.

Peroxidase-Antiperoxidase (PAP) Method

The PAP is an unlabeled antibody method first developed by Sternberger and his associates (74) and used for the detection of antitreponemal antibodies. The PAP reagent is an antibody against horseradish peroxidase and horseradish peroxidase antigen in the form of a small stable immune complex consisting of two antibody molecules and three horseradish peroxidase molecules (74). There technique is much more sensitive than the indirect technique.

Avidin-Biotin Complex Peroxidase (ABC) Method

This technique was developed by Hsu and his colleagues for immunostaining tissue sections. There is high-affinity binding between avidin and biotin, which are used to detect the primary antibody (32,33). Avidin is chemically conjugated to horseradish peroxidase and there is high binding between the avidin and biotin with subsequent localization of the peroxidase by the reaction with the chromogen. The ABC procedure has been found to be even more sensitive than the PAP method (32,33). Endogenous biotin that is present in some tissues can be blocked with various techniques such as preincubation with avidin and biotin (76).

Additional Methods

Various other immunocytochemical methods have been used to detect antigens in tissues. These include the unlabeled antibody method with the enzyme bridge technique, the protein A-peroxidase conjugated method, the protein A-PAP method, and the enzyme-labeled antigen method. These techniques are discussed in more detail in recent publications (12,13,58,60,72,76).

Enzyme System

Various enzyme systems have been used to detect antigens in immunocytochemical reactions by producing a visible color change in a substrate at the sites of the antibody-enzyme complex. Horseradish peroxidase is used most commonly, but other enzyme systems including alkaline phosphatase, glucose oxidase, and β-galactosidase have been successfully used to localize antigen-antibody reactions in endocrine tissue.

Protein A

Protein A is present in the cell wall of the bacterium *Staphylococcus aureus*. It can attach to the Fc portion of immunoglobulin molecules. Protein A can be labeled with various markers, including colloidal gold particles and fluorescent tags, and can be used to localize antigens at both the light and electron microscope levels (1,63). The affinity of protein A for immunoglobulin varies from one species to another and with the subtype of immunoglobulin, so binding affinity must be evaluated for each species and subtype of immunoglobulin.

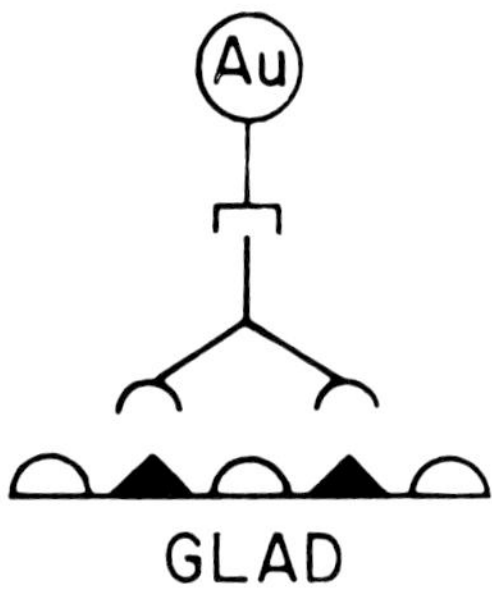

Figure I.2. Gold-labeled antigen detection (GLAD) technique can be used at the light or electron microscope level to detect specific antigens. After incubation with gold particles linked to strepavidin or to specific immunoglobulins, the colloidal gold particles can be detected by light or electron microscopy.

Immunofluorescence

Immunofluorescence microscopy can be used to localize antigens in cells. It has been in use longer than enzyme immunochemistry. Antibodies can be linked directly or indirectly to fluorescein or rhodamine isothiocyanate, which will give a distinct pattern of fluorescence at specific wavelengths. The bright apple-green fluorescence of fluorescein isothiocyanate with excitation at 490 nm and emission at 520 nm can be detected with a fluorescence microscope. The use of fluorescence techniques in immunocytochemistry has some advantages, including a) the speed and simplicity of this approach and b) the excellent resolution provided. However, the disadvantages of the fluorescence methods have made this technique less attractive in diagnostic endocrine pathology. The main disadvantages include a) the necessity of using a fluorescence microscope, b) the difficulties in interpreting background details, and c) the fact that the fluorescent preparations are not permanent. A unique feature of fluorescence microscopy that may cause problems in interpretation of results is autofluorescence. Autofluorescence can be seen with many tissues. It is usually a yellow or bluish color and can be recognized when the specific green fluorescence is very strong. Distinguishing autofluorescent background from specific fluorescence is often difficult with weakly stained specimens.

Immunogold Labeling

Protein A coupled with colloidal gold can be used for immunochemical studies at the light microscope and electron microscope levels (Figs. I.2, I.3, I.4). The sensitivity of protein A-gold labeling can be increased by combining it with the silver staining technique (31). Dilute immunogold is applied to tissue sections after the primary antibody, and the gold particles that are localized at antigenic sites are revealed under direct light microscopy by a silver precipitation reaction. The advantages of this method include 1) ease of use in paraffin sections and 2) extreme sensitivity. It can also be used for double or triple immunostaining procedures (25,31). Roth has employed protein A-gold techniques with gold and silver complexes to localize single and multiple antigens at the light microscope level in paraffin sections (62).

Multiple Antigen Localization

When it is necessary to localize two or more antigens within the same cell or tissue section, this may be done by a) using two or more different enzyme systems, b) using two or more chromogens with the same enzyme system, and c) combining the foregoing procedures with immunogold or with immunofluorescence labeling. Nakane first showed that it was possible to localize two antigens at the light microscopic level by the elution technique (52). This approach worked well at the light microscope level if the antigens of interest were localized in different cellular compartments and if they were not destroyed by the elution technique. Recent studies have utilized double and triple staining without elution (18,79). Multiple labeling techniques are most helpful when the antigens are present in different cell populations or in nonoverlapping populations

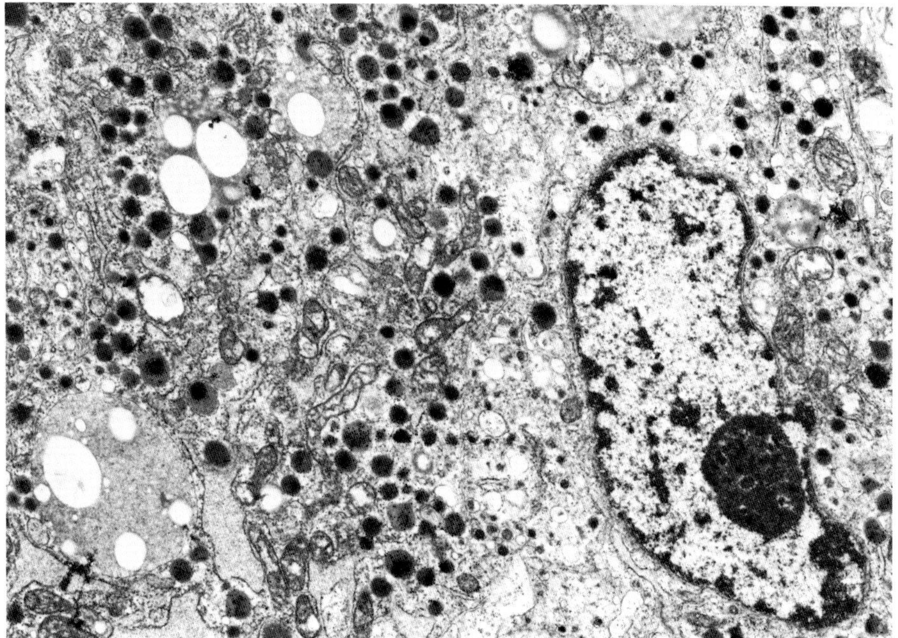

Figure I.3. Localization of prolactin hormone in pituitary cells by colloidal gold labeling with 15-nm gold particles (×9804).

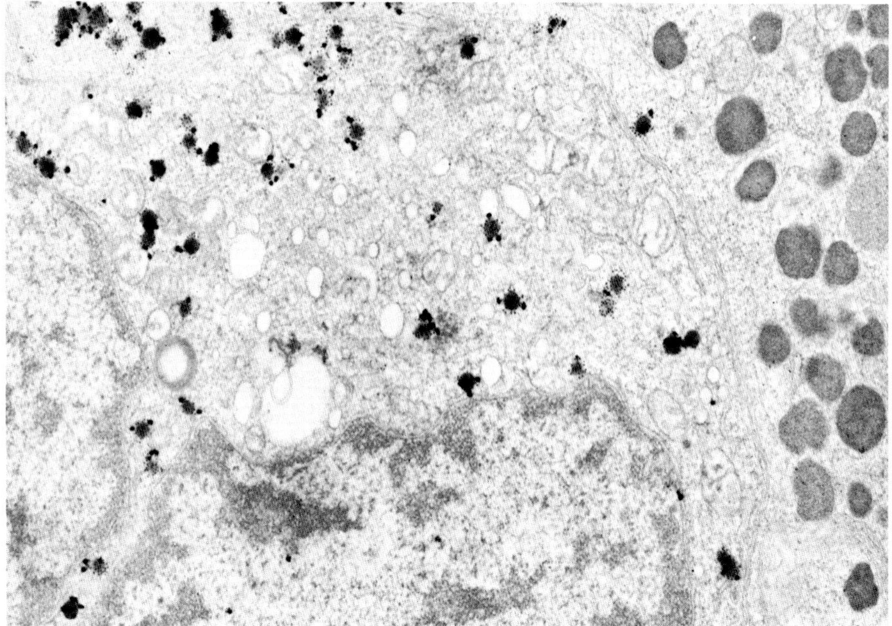

Figure I.4. Localization of both GH (40-nm gold particles) and PRL (15-nm gold particle) in a pituitary cell (×15,000).

and when it is necessary to know the codistribution of two or more cell populations.

Applications of Immunocytochemistry in Endocrine Pathology

A wide variety of applications of immunocytochemical techniques in endocrine pathology have been described. These have ranged from studies of multiple hormone expression in complex endocrine glands to the analysis of receptor proteins in target tissues. Broad-spectrum neuroendocrine markers have been used to identify cells of the DNS at the light microscope level. These markers include neuron-specific enolase, chromogranins, synaptophysin, bombesin/gastrin releasing peptide and Leu 7 (HNK-1) (Table I.1) (2,5–7,22,23,26, 34,45–47,54–56, 61, 75, 77, 80–82).

Table I.1. Immunochemical markers used in the study of selected endocrine cells and tumors

Cell/neoplasm	Endocrine markers[a]
Adrenal medulla and paraganglia	Chromogranins, neuron-specific enolase, synaptophysin, enkephalins, proopiomelanocortin (15,21,23, 44,47,48,54,68,82)[b]
Adrenal cortex	Cytochrome P450-21 hydroxylase (67a)
	Cytochrome $P450_{17\alpha}$ (67)
Gastrointestinal endocrine	Bombesin, somatostatin, gastrin, serontonin, cholecystokinin, secretin (11,21,23,45,47)
Hypothalamus	GHRH, GnRH, somatostatin, TRH, CRH, vasopressin, oxytocin (3,41,43,57)
Lung endocrine	ACTH, bombesin, Leu 7, calcitonin, enkephalins, CRH (7,21,27,68,80)
Merkel cells	Chromogranins, calcitonin, VIP, synaptophysin (20,69,82)
Pancreatic endocrine	Insulin, glucagon, somatostatin, pancreatic polypeptide, VIP, gastrin, ACTH (8,11,16,21,23,28,70)
Pituitary	GH, PRL, ACTH, FSH/LH, TSH, alpha subunit, chromogranins, synaptophysin (21,30,39,40,61,70)
Parathyroid	Parathyroid hormone, parathyroid hormone-related peptide, chromogranins (17,82)
Thyroid C cells	Calcitonin, calcitonin gene-related peptide, katacalcin, somatostatin (14,42, 49,51,65,70,78)
Thyroid follicular cells	Thyroglobulin (4)

[a]Abbreviations: GHRH, growth hormone releasing hormone; GnRH, gonadotropin releasing hormone; TRH, thyrotropin releasing hormone; CRH, corticotropin releasing hormone; VIP, vasoactive intestinal polypeptide.
[b]Reference numbers in parentheses.

Neuroendocrine Markers That May Have Prognostic Significance in Endocrine Diseases

The use of immunohistochemical markers has been reported to be helpful in predicting the biological behavior of selected endocrine neoplasms including medullary thyroid carcinomas, pancreatic endocrine tumors, and hormone-responsive tumors such as breast carcinomas.

Medullary Thyroid Carcinomas

The C cells of the thyroid produce calcitonin and many other peptides. The immunochemical distinction between C-cell hyperplasia and medullary thyroid carcinoma (MTC) was made by histaminase immunostaining by Mendelsohn and his colleagues (50). This enzyme was reported to be present in MTC but not in normal or hyperplastic C cells (50). Although calcitonin immunoreactivity is present in normal and hyperplastic C cells as well as in MTC, rare tumors may not contain immunoreactive calcitonin (78). Many other substances, including glucagon, gastrin, insulin, gastrin releasing peptide, human chorionic gonadotropin, serontonin, catecholamines, and carcinoembryonic antigen (CEA) have been found in MTC (14,42,49). Immunohistochemical studies have shown that some cases of "anaplastic" thyroid carcinomas have calcitonin and are really medullary thyroid carcinomas (49). Several recent studies have suggested that patients with calcitonin-poor tumors have a worse prognosis than patients with calcitonin-rich tumors and that the loss of calcitonin expression, with increased expression of CEA, may indicate a more aggressive biological behavior (51,64,65). These findings have not yet been extensively corroborated.

and unlabeled antibody (PAP) procedures. J Histochem Cytochem 29:577–580

34. Iwase K, Nagasaka A, Kato K, Ohtani S, Nagatsu I, Ohyama T, Kakai A, et al. (1986) Enolase subunits in patients with neuroendocrine tumors. J Clin Endocrinol Metab 63:94–101
35. Kahn CR, Rosen SW, Weintraub BD, Fajans SS, Gorden P (1977) Ectopic production of chorionic gonadotropin and its subunits by islet-cell tumors. A specific marker for malignancy. N Engl J Med 297:565–569
36. King WJ, DeSombre ER, Jensen EV, Greene GL (1985) Comparison of immunocytochemical and steroid-binding assays for estrogen receptor in human breast cancer tumors. Cancer Res 45:293–304
37. King WJ, Greene GL (1984) Monoclonal antibodies localize oestrogen receptor in the nuclei of target cells. Nature 307:745–747
38. Kohler G, Milstein C (1975) Continuous cultures of fused cells secreting antibody of predefined specificity. Nature 256:495–497
39. Kovacs K, Horvath E (1986) Tumors of the Pituitary Gland. Atlas of Tumor Pathology, Second Series, Fascicle 21. Washington, DC, Armed Forces Institute of Pathology
40. Kovacs K, Horvath E, Corenblum B, Sirek AMT, Penz G, Ezrin C (1975) Pituitary chromophobe adenomas consisting of prolactin cells. A histologic immunocytological and electron microscopic study. Virchows Arch [Pathol Anat] 366:113–123
41. Krisch B (1978) Hypothalamic and extrahypothalamic distribution of somatostatin-immunoreactive elements in the rat brain. Cell Tissue Res 195:499–513
42. Krisch K, Krisch I, Horvat G, Neuhold N, Ulrich W (1985) The value of immunohistochemistry in medullary thyroid carcinoma: a systematic study of 30 cases. Histopathology 9:1077–1089
43. Lechan RM, Jackson IMD (1983) Immunohistochemical localization of thyrotropin-releasing hormone (TRH) in the rat hypothalamus and pituitary. Endocrinology 111:55–65
44. Linnoila RI, Lack EE, Steinberg SM, Keiser HR (1988) Decreased expression of neuropeptides in malignant paragangliomas: An immunohistochemical study. Hum Pathol 19:41–50
45. Lloyd RV, Mervak T, Schmidt K, Warner TFCS, Wilson BS (1984) Immunohistochemical detection of chromogranin and neuron-specific enolase in pancreatic endocrine neoplasm. Am J Surg Pathol 8:607–614
46. Lloyd RV, Sisson JC, Shapiro B, Verhofstad AAJ (1986) Immunohistochemical localization of epinephrine, norepinephrine, catecholamine-synthesizing enzymes, and chromogranin in neuroendocrine cells and tumors. Am J Pathol 125: 45–54
47. Lloyd RV, Warner TF (1984) Immunohistochemistry of neuron-specific enolase. In: DeLellis RA (ed) Advances in Immunohistochemistry. New York, Masson, pp 127–140
48. Lloyd RV, Wilson BS (1983) Specific endocrine marker defined by a monoclonal antibody. Science 222:628–630
49. Mendelsohn G, Bigner SH, Eggleston JC, Baylin SB, Wells SA (1980) Anaplastic variants of medullary thyroid carcinoma. A light microscopic and immunohistochemical study. Am J Surg Pathol 4: 333–341
50. Mendelsohn G, Eggleston JC, Weisburger WR, Gann DS, Baylin SB (1978) Calcitonin and histaminase in C-cell hyperplasia and medullary thyroid carcinoma. A light microscopic and immunohistochemical study. Am J Pathol 92:35–52
51. Mendelsohn G, Wells SA, Baylin SB (1984) Relationship of tissue carcinoembryonic antigen and calcitonin in tumor virulence in medullary thyroid carcinoma. An immunohistochemical study in early, localized, and virulent disseminated stages of disease. Cancer 54:657–662
52. Nakane PK (1968) Simultaneous localization of multiple tissue antigens using the peroxidase-labeled antibody method: a study on pituitary glands of the rat. J Histochem Cytochem 16: 557–560
53. Nakane PK, Pierce GB Jr (1966) Enzyme-labeled antibodies: preparation and application for the localization of antigen. J Histochem Cytochem 14: 929–931
54. O'Connor DT (1983) Chromogranin: widespread immunoreactivity in polypeptide hormone producing tissues and in serum. Regul Peptide 6: 263–280
55. O'Connor DT, Frigon RP (1984) Chromogranin A, the major catecholamine storage vesicle soluble protein. Multiple size forms, subcellular storage, and regional distribution in chromaffin and nervous tissue elucidated by radioimmunoassay. J Biol Chem 259:3237–3247
56. Pahlman S, Esscher T, Nilsson K (1986) Expression of γ-subunit of enolase, neuron specific enolase, in human non-neuroendocrine tumors and derived cell lines. Lab Invest 54:554–560
57. Pelletier G (1979) Immunohistochemical localization of hypothalamic hormones and other peptides in the central nervous system. In: Collu R, Barbeau A, Ducharme JR, et al. (eds) The Central Nervous System. Effects of Hypothalamic Hormones. New York, Raven, pp 331–334
58. Polak JM, Van Noorden (eds) (1986) Immunocyto-

chemistry. Modern Methods and Applications. Bristol, Wright
59. Polak JM, Bloom SR, Hobbs S, Solcia E, Pearse AGE (1985) Distribution of a bombesin-like peptide in human gastrointestinal tract. Lancet 1: 1109–1110
60. Polak JM, Varndell IM (1984) Immunolabeling for Electron Microscopy. Amsterdam, Elsevier
61. Rosa P, Fumagalli G, Zanini A, Huttner WB (1985) The major tyrosine sulfated protein of the bovine anterior pituitary is a secretory protein present in gonadotrophs, thyrotrophs, mammotrophs and corticotrophs. J Cell Biol 100:928–937
62. Roth J (1982) Applications of immunocolloids to light microscopy: Preparation of protein A-silver and protein A-gold complexes and their application for localization of single and multiple antigens in paraffin sections. J Histochem Cytochem 30:691–696
63. Roth J, Bendayan M, Orci L (1978) Ultrastructural localization of intracellular antigens by the use of protein A-gold complex. J Histochem Cytochem 26:1074–1081
64. Ruppert JM, Eggleston JC, de Bustros A, Baylin SB (1986) Disseminated calcintonin-poor medullary thyroid carcinoma in a patient with calcitonin-rich primary tumor. Am J Surg Pathol 10:513–518
65. Saad MF, Ordonez NG, Guido JJ, Samaan NA (1984) The prognostic value of calcitonin immunostaining in medullary carcinoma of the thyroid. J Clin Endocrinol Metab 59:850–856
66. Sar M, Parikh I (1986) Immunohistochemical localization of estrogen receptor in rat brain, pituitary and uterus with monoclonal antibodies. J Steroid Biochem 24:497–503
67. Sasano H, Mason JI, Sasano N (1989) Immunohistochemical study of cytochrome P-$450_{17\alpha}$ in human adrenocortical disorders. Hum Pathol 20: 113–117
67a. Sasano H, White PC, New MI, Sasano N (1988) Immunohistochemical localization of cytochrome P-$450_{c\text{-}21}$ in human adrenal cortex and its relation to endocrine function. Hum Pathol 19:181–185
68. Schmechel D, Marangos PJ, Brightman M (1978) Neuron-specific enolase is a molecular marker for peripheral and central neuroendocrine cells. Nature 276:834–836
69. Sibley RK, Dahl D (1985) Primary neuroendocrine (Merkel cell?) carcinoma of the skin. II. An immunocytochemical study of 21 cases. Am J Surg Pathol 9:109–116
70. Sobol RE, Memoli V, Deftos LJ (1989) Hormone-negative, chromogranin A-positive endocrine tumors. N Engl J Med 320:444–447
71. Stefanini P, Carboni M, Patrassi H, Basoli A (1974) Beta-islet cell tumor of the pancreas: Results of a study of 1067 cases. Surgery 75:597–609
72. Sternberger LA (1986) Immunocytochemistry, 2nd ed. New York, Wiley
73. Sternberger LA (1979) Immunocytochemistry, 2nd ed. New York, Wiley
74. Sternberger LA, Hardy PH Jr, Cuculis JJ, Meyer HG (1970) The unlabeled antibody-enzyme method of immunohistochemitry. Preparation and properties of soluble antigen-antibody complex (horseradish peroxidase-antihorseradish peroxidase) and its use in the identification of spirochetes. J Histochem Cytochem 18:315–333
75. Tatemoto K, Efendic S, Mutt V, Makk G, Feistner GJ, Barchas JD (1986) Pancreastatin, a novel pancreatic peptide that inhibits insulin secretion. Nature 324:476–478
76. Taylor CR (1986) Immunomicroscopy: a diagnostic tool for the surgical pathologist. Philadelphia, Saunders
77. Tischler AS, Mobtaker H, Mann K, Nunnemacher G, Jason WJ, Dayal Y, DeLellis RA, Adelman L, Wolfe HJ (1986) Anti-lymphocyte antibody Leu-7 (HNK-1) recognizes a constituent of neuroendocrine granule matrix. J Histochem Cytochem 34: 1213–1216
78. Uribe M, Grimes M, Fenoglio-Preiser CM, Feind C (1985) Medullary carcinoma of the thyroid gland. Clinical, pathological, and immunohistochemical features with review of the literature. Am J Surg Pathol 9:577–594
79. Wang B-L, Larsson L-I (1985) Simultaneous demonstration of multiple antigens by indirect immunofluorescence or immunogold staining. Novel light and electron microscopical double and triple staining method employing primary antibodies from the same species. Histochemistry 83:47–56
80. Wharton J, Polak JM, Bloom SR, Ghatei MA, Solcia E, Brown MR, Pearse AGE (1978) Bombesin-like immunoreactivity in the lung. Nature 273: 769–770
81. Wiedenmann B, Franke WW, Kuhn C, Moll R, Gould VE (1986) Synaptophysin: a marker protein for neuroendocrine cells and neoplasms. Proc Natl Acad Sci USA 83:3500–3504
82. Wilson BS, Lloyd RV (1984) Detection of chromogranin in neuroendocrine cells with a monoclonal antibody. Am J Pathol 115:458–468
83. Woodruff JM, Huvos AG, Erlandson RA, Shah JP, Gerold FP (1985) Neuroendocrine cracinomas of the larynx. A study of two types, one of which mimics thyroid medullary carcinoma. Am J Surg Pathol 9:771–790

Avidin-Biotin Complex Immunoperoxidase Method

Procedure

1. Deparaffinize slide using xylene, two changes each 10 min.
2. Wash in absolute ethanol 4 min, two changes.
3. Wash in 95% ethanol 3 min, two changes.
4. Rinse in distilled water.
5. Treat with (3%) hydrogen peroxide in absolute methanol 10 min.
6. Wash three times in phosphate-buffered saline (PBS) 2 min each.
7. Incubate in diluted normal serum 30 min (suppressor serum). Normal goat serum, 1:20 dilution (e.g., 0.5 ml serum + 9.5 ml NGS/BSA/PBS).
8. "Shake off" serum and directly add primary antiserum.

Note: Dilutions of the primary antiserum depend on the particular system one is using. Timing is also different and may vary from a few minutes to an overnight incubation.

9. Wash in PBS three times, 2 min each.
10. Incubate in biotinylated anti-rabbit IgG made in goat for 30 min. Dilution—biotinylated 10µl IgG + 2 ml of diluent (NGS/BSA/PBS).
11. Wash in PBS, three changes, 2 min each.
12. Incubate in ABC reagent 30 min. Dilution 25µl of reagent A, 25µl of reagent B, and 2.5 ml of diluent (NGS/BSA/PBS).
13. Rinse in PBS, two changes, 2 min each.
14. Place in DAB-PBS for staining 10 min. (To 25 ml PBS add 10 mg diaminobenzidine* and filter; to this add 5 µl of 30% or 50 µl of 3% hydrogen peroxide.)
15. Wash in H_2O two times, 5 min each.
16. Wash in distilled water and, if desired, counterstain.
17. Dehydrate, clear, and mount.

Note: Diluent—NGS/BSA/PBS—2% bovine serum albumin in PBS; to this add normal goat serum to make a 1% solution. 1% NGS/4% BSA/PBS—make 100 ml for stock solution. Keep refrigerated or frozen.

Avidin-Biotin Complex Immunoperoxidase Method for Frozen Sections

Preparation of Tissues

Freeze tissues with OCT in liquid nitrogen. Store at −70°C. (Antigens are stable for many months with this method.)

Cut sections at 8 µm; dry sections briefly and then fix with cold reagent grade acetone for 10 min. Allow sections to dry and then freeze, preferably at −70°C.

Staining Procedure

1. Thaw tissue sections on paper towel until dry—5 to 10 min.
2. Refix with reagent grade acetone for 1 min (optional).
3. Allow sections to dry for 5 min.
4. Place sections in PBS for 5 min.
5. Incubate in diluted normal serum (suppressor serum) 1:20 for 10 min.
6. Shake off suppressor serum and add primary antiserum. Timing and dilutions of the primary antiserum will depend on the specific antiserum being used.
7. Wash in PBS three times, 2 min each.
8. Incubate slides with biotinylated IgG for 30 min. Use 1:200 dilution (e.g., 10λ biotinylted IgG + 2 ml diluent).
9. Wash in PBS three times, 2 min each.
10. Apply 3% hydrogen peroxide–absolute methnol for 10 min (equals parts).
11. Wash in PBS three times, 2 min each.
12. Incubate in ABC reagent, 30 min: 25λ A + 25λ reagent B and 2.5 ml of diluent (NGS/BSA/PBS).
13. Wash in PBS three times, 2 min each.
14. Apply PBS-DAB for 10 min. Refer to preced-

---

*DAB is a potential carcinogen. Add 4 ml of PBS with a syringe and needle through the preweighed 10-mg vial of DAB. Allow to dissolve for 30 to 60 min with periodic shaking or sonicale for 1 min. After DAB is in solution add 21 ml of PBS. Preweighed DAB can be obtained commercially (Sigma).

ing section for DAB* "recipe" and precautions.

15. Rinse sections in two changes of distilled water for 5 min each.
16. Counterstain with hematoxylin for 1 min.
17. Dehydrate and mount.

Postembedding Ultrastructural Immunohistochemical Procedure Using ABC Method

1. Fixation: Fixation will vary with specific antigen and tissue under study, requiring some degree of preliminary study of suitability. In general, a fixative that best preserves the specific antigen as well as the ultrastructure of the cell is recommended. We use 4% paraformaldehyde and 1% glutaraldehyde, pH 7.2, for 1 hour. Osmium is *not* used as a counterstain.
2. Wash in buffer.
3. Dehydration: Dehydration should be performed with increasing concentrations of ethanol. The use of propylene oxide as a clearing agent is controversial.
4. Embedding: Embedding is done in Epon or Araldite and Polybed.
5. Sectioning: 0.5–1 μm sections are taken for tissue orientation and evaluation for immunoperoxidase reaction for the specific antigen at the light microscopic level after etching. Failure to demonstrate positive staining in thick sections does not exclude, but makes less likely, successful ultrastructural localizations.
6. Ultrathin sectioning: Collect thin sections on nickel grids. It is unnecessary to use coated grids as sections usually don't fall off or wrinkle.
7. Etching (optional step): Float grid section side down on droplet of 5% H_2O_2 for 10 min.
8. Rinse well with filtered water from squirt bottle.
9. Float the grids on drops of PBS.
10. Method for immunostaining of grids: All grids are stained by placing them on drops of appropriately diluted antisera on a clean glass slide.
11. Immunostaining protocol:
    A. Blocking step—5% suppressor serum for 10 min.
    B. A drop of primary antiserum is placed on glass slide as above. Grids are floated for 1 hour at room temperature or overnight in the refrigerator in a moist chamber. Refrigerated sections are allowed to come to room temperature for an hour before proceeding.
    C. Sections are washed well with buffer and treated with biotinylated IgG for 30 min.
    D. Grids are rinsed well and treated with ABC for 30 min.
    E. React with DAB for 5 min.
    F. Wash with distilled water.
    G. Counterstain with 5% aqueous uranyl acetate for 30 min.

Postembedding Ultrastructural Immunohistochemical Procedure Using the Gold Method

1. Tissues and grids are handled similarly to the ABC method.
2. Blocking step: Grids are rinsed in buffer and inverted on drops of 1% bovine albumin for 5 min.
3. They are then plotted, not rinsed, and placed on drops of primary antiserum.
4. Grids are incubated on the primary antiserum for 1 to 2 hours at room temperature in a moist chamber.
5. Two short washes of PBS.
6. Grids are placed on drops of commercially available colloidal gold.
7. Grids are washed well with buffer and then rinsed in double distilled water.
8. Grids are stained using 5% aqueous uranyl acetate for 30 min.

*DAB (diaminobenzidine) is a potential carcinogen.

Applications of In Situ Hybridization in Endocrine Pathology

Use of Broad-Spectrum Probes

Many broad-spectrum neuroendocrine markers, including the chromogranins, synaptophysin, protein 7B2, neuron-specific enolase, and protein gene product 9.5 (PGP 9.5), can be used to characterize neuroendocrine cells by immunochemistry (9). With the recent advances in molecular endocrine biology, the cDNA clones and derived mRNA and protein sequences for many hormone gene products have also been analyzed. Such studies have shown that there is usually one mRNA species that may be processed into various protein products by proteolytic cleavage of prohormone in the cytoplasm. Examples are the mRNA for proopiomelanocortin, which is processed to ACTH, β-endorphin, melanocyte-stimulating hormone, and several other biologically active peptides (67). Likewise, the pancreatic glucagon precursor preproglucagon, which is processed to glucagon, glicentin or enteroglucagon, and various other glucagon-related peptides (27).

The use of one probe to detect the intracellular messenger RNA before posttranslational processing occurs is a rapid means of characterizing endocrine cells of interest.

The chromogranins are a complex family of proteins with the predominant forms known as chromogranins A, B, and C (26,32). However, several chromogranin A and B peptides are derived from the proteolytic cleavage of the parent molecules, and these cleavage products have variable distribution in different tissues. Examples are pancreastatin and betagranin from chromogranin A and various other peptides from chromogranin B (2,3,4,40– 42,59,60,69). Chromogranin A is an important marker in diagnostic pathology because it helps to characterize neuroendocrine cells (59,60,69). For diagnostic purposes one can use various antibodies against different portions of the chromogranin A molecule such as pancreastatin and betagranin. Alternatively, one can use ISH to detect the intact mRNA for chromogranin A that contains the sequence for both pancreastatin and betagranin. This mRNA is present in most endocrine tissues with secretory granules (57). For experimental and analytical purposes, specific antibodies against various fragments of the chromogranins or nucleic acid probes for specific chromogranins can be extremely useful.

In situ hybridization studies with probes for chromogranins A and B have shown a widespread distribution of these mRNAs (57). In addition, many neoplasms with few secretory granules that express only small amounts of immunoreactive chromogranin A, such as small-cell lung carcinoma, usually contain abundant mRNAs for chromogranin A (57).

Oncogene and Growth Factor Expression in Endocrine Pathology

Studies of oncogene expression in endocrine neoplasms with hybridization analysis have been reported by many investigators (16,18,52, 66). Studies of N-*myc* gene expression in neuroblastomas have determined that high levels of expression of this oncogene are related to a poor prognosis (18). Many oncogenes have been reported to be altered in lung carcinomas, including *myc, ras* and *erb-B-2*. Alterations in *erb-B-1* and *erb-B-2* appeared to correlate with histological types of lung tumors and were most common in metastatic tumors (16). Activated *ras* oncogene has been noted in many thyroid carcinomas. In one study, 80% of follicular neoplasms but only 20% of papillary neoplasms showed activated *ras* genes (52). Analysis of the expression of various oncogenes by medullary thyroid carcinomas showed that these tumors expressed c-*myc* and Ha-*ras* mRNA (35). Hypothalamic growth hormone releasing factor (GRF) has been found to induce c-*fos* expression in cultured pituitary cells, suggesting that GRF may be involved in events leading to growth hormone cell transformation (7). GH was noted to induce c-*myc* expression in the liver and kidney (66).

Growth factors play an essential role in regulating cellular function in normal and neoplastic endocrine cells (74). Insulinlike growth factor I simulates growth hormone mRNA in transplantable pituitary tumors (22). Bombesin stimulated prolactin hormone secretion from cultured rat pituitary tumor cells (10). These findings implicate growth factors in basic regulatory mechanisms in endocrine neoplasms. Specific growth factors such as epidermal growth factor may be involved in activation of specific *ras* oncogene proteins (49).

Hypothalamus

Reports on the uses of ISH in many areas of endocrine pathology have appeared (24,33,58,68,71, 75,85,89). Vasopressin gene expression in the hypothalamus was shown to vary with dehydration and salt loading in the rat by Sherman et al. (75). Uhl et al. also studied vasopressin gene expression in the hypothalamus (85). The distribution of somatostatin mRNA in the rat hypothalamus has been analyzed by Hofler et al. (33) and confirmed earlier immunocytochemical analyses on the wide distribution of this peptide.

Pituitary

In situ hybridization studies have been used to study normal and neoplastic pituitary tissues in humans and other species (55,56,58,68,71,89). Studies of estrogen regulation of rat prolactin messenger RNA levels have provided morphological evidence of increased prolactin gene levels in specific cell types with estrogen treatment (58, 68,89). In some cases the translated products were more difficult to detect with specific antibodies (56).

Hybridization studies with human pituitary adenomas have shown that nonfunctional (null cell) tumors commonly produced the beta subunit of human chorionic gonadotropin (44,45). These tumors also produce glycoprotein hormone mRNAs, including FSH and LH gene products (44,45). Another study (73a) revealed PRL and ACTH mRNAs in null cell adenomas, which were not readily appreciated by immunochemistry.

Analysis of growth hormone- and prolactin-secreting pituitary adenomas by Lloyd and co-workers revealed that a high percentage of tumors causing acromegaly also expressed prolactin mRNA (55,56). Lloyd and co-workers found a pituitary adenoma secreting large amounts of growth hormone. This tumor did not store detectable growth hormone and was negative for growth hormone by immunostaining. However, the mRNA for growth hormone was readily detected in the tumor by ISH (56). Other ISH studies have shown that some tumors that are clinically silent and do not produce acromegaly express the gene for growth hormone (50), suggesting a defect in transcription or a dysfunctional translation product.

Thyroid Gland

The expression of thyroglobulin mRNA has been studied extensively in thyroid neoplasms by Berge-Lefranc and colleagues (5). They found that well-differentiated papillary and follicular neoplasms expressed the same levels of thyroglobulin mRNA per cell, while moderately differentiated carcinomas contained less thyroglobulin mRNA. These findings suggested that there might be a relationship between the degree of differentiation and mRNA expression. However, many more studies are needed to evaluate these preliminary findings. Calcitonin and calcitonin gene-related peptide (CGRP) have been found in the same cells by ISH in medullary thyroid carcinomas (MTC) (37,90, 91). The calcitonin gene has been detected in all cases of MTC in various hybridization studies. The proopiomelanocortin (POMC) gene was also detected in MTC by various investigators (37,80). A frequent observation has been that the mRNA for POMC in MTC and other neoplasms outside the pituitary is usually 100 to 250 bases larger than the pituitary gene product (20). The significance of this observation is still unknown. Hofler and colleagues have localized both calcitonin mRNA and peptide in MTC (36).

Adrenal Glands

Pheochromocytomas and paragangliomas are known to produce multiple peptides and amines. The mRNAs for various peptides have been localized in pheochromocytomas (37). Other workers have found the mRNAs for calcitonin, gastrin-releasing peptide, and POMC in pheochromocytomas (37,86). Haselbacher and co-workers found that insulinlike growth factor II mRNA was expressed in pheochromocytomas (31). The tumors studied contained 20 times more IGF-II protein than normal adrenals (31). Chromogranin A and B mRNAs have been detected in the normal adrenal medulla and in pheochromocytomas (57).

Ectopic Hormone Production

The hypercalcemia present in patients with malignant neoplasms may have different etiologies. A recent study of patients with hypercalcemia associated with malignant disease revealed that expression of parathyroid gene was not related to this condition (77). Several investigators have now cloned the parathyroid hormone-related gene responsible for many cases of malignancy-associated hypercalcemia (43,82). Parathyroid hormone-related peptide is expressed in many nonendocrine tissues, especially squamous cell carcinomas, but is also found in many endocrine tissues and tumors including parathyroid adenomas (82).

Ectopic hormone production has been characterized by gene product expression in various neoplasms, including GH mRNA production in a pancreatic endocrine tumor producing acromegaly (62) and proopiomelanocortin products in an islet cell carcinoma of the pancreas causing Cushing's syndrome (61). Molecular analysis of the gene products in a leiomyosarcoma from a patient with hypoglycemia revealed that the tumor produced IGF-II. IGF-II also appeared to inhibit growth hormone secretion in the patient (19a). Other ISH analyses have produced surprising results such as the production of proopiomelanocortin by testicular Leydig cells (steroid-producing cells) and of calcitonin gene-related peptide by cultured Ewing's sarcoma cells (38,70). In situ hybridization studies have been used to detect gene products for pancreatic hormones in pancreatic endocrine tumors, including the localization of pancreatic polypeptide and vasoactive intestinal polypeptide gene products in cells that were not expressing the translated protein products (21). The localization of gene products of ectopic hormones in endocrine cells and tumors reported by several investigators, including proopiomelanocortin messenger RNA in the adult rat testis (70) and angiotensinogen in rat brain (81), indicates how useful hybridization histochemistry has become in characterizing specific gene products in endocrine cells.

The use of in situ hybridization to study endocrine lesions is still in its infancy. Many major advances and the development of more sensitive techniques can be anticipated in the near future.

References

1. Alwine JC, Kemp DJ, Parker BA, Reiser J, Renart J, Stark GR, Wahl GM (1979) Detection of specific RNAs or specific fragments of DNA by fractionation in gels and transfer to diazobenzyloxymethyl paper. Methods Enzymol 68:220–242
2. Benedum UM, Baeuerle PA, Konecki DS, Frank R, Powell J, Mallet J, Huttner WB (1986) The primary structure of bovine chromogranin A: a representative of a class of acidic secretory proteins common to a variety of peptidergic cells. EMBO J 5:1495–1502
3. Benedum UM, Lamouroux A, Konecki DS, Rosa P, Hille A, Baeuerle PA, Frank R, Lottspeich F, Mallet J, Huttner WB (1987) The primary structure of human secretogranin I (chromogranin B): comparison with chromogranin A reveals homologous terminal domains and a large intervening variable region. EMBO J 6:1203–1211
4. Benjannet S, Leduc R, Adrouche N, Falgueyret JP, Marcinkiewicz M, Seidah NG, Mbikay M, Lazure C, Chretien M (1987) Chromogranin B (secretogranin I), a putative precursor of two novel pituitary peptides through processing at paired basic residues. FEBS Lett 224:142–148
5. Berge-Lefranc JL, Cartouzou G, De Micco C, Fragu P, Lissitzky S (1985) Quantification of thyroglobulin messenger RNA by in situ hybridization of differentiated thyroid cancers. Difference between well differentiated and moderately differentiated histologic types. Cancer 56:345–350
6. Bhatnagar M, Springall DR, Ghatei MA, Burnet PWJ, Hamid Q, Giaid A, Ibrahim NBN, Cuttita F, Spindel ER, Penketh R, Rodek C, Bloom SR, Polak JM (1988) Localisation of mRNA and coexpression and molecular forms of GRP gene products in endocrine cells of fetal human lung. Histochemistry 90:299–307
7. Billestrup N, Mitchell RL, Vale W, Verma IM (1987) Growth hormone releasing factor induces c-*fos* expression in cultured primary pituitary cells. Mol Endocrinol 1:300–305
8. Binder M, Tourmente S, Roth J, Renaud M, Gehring WJ (1986) In situ hybridization at the electron microscope level: localization of transcripts on ultrathin section of Lowicryl K4M-embedded tissue using biotinylated probes and protein A-gold complexes. J Cell Biol 102:1646–1653
9. Bishop AE, Power RF, Polak JM (1988) Markers for neuroendocrine differentiation. Pathol Res Pract 183:119–128
10. Bjoro T, Torjesen PA, Ostberg BC, Sand O, Iversen J-G, Gautvik KM, Haug E (1987) Bombesin stimulates prolactin secretion from cultured rat pituitary tumor cells (GH_4C_1) via activation of phospholipase C. Regul Peptides 19:169–182
11. Bloch B, Popovici T, LeGuellec D, Normand E, Chouham S, Guitteny AF, Bohlen P (1986) In situ hybridization histochemistry for the analysis of

gene expression in the endocrine and central nervous system tissues: A 3-year experience. J Neurosci Res 16:183–200

12. Brahic M, Haase AT (1978) Detection of viral sequences of low reiteration frequency by in situ hybridization. Proc Natl Acad Sci USA 75:6125–6129
13. Brigati DJ, Meyerson D, Leary JJ, Spalholz B, Travis SZ, Fong CKY, Hsiung GD, Ward DC (1983) Detection of viral genomes in cultured cells and paraffin embedded tissue sections using biotin-labeled hybridization probes. Virology 126: 32–50
14. Caskey CT (1987) Disease diagnosis by recombinant DNA methods. Science 236:1223–1229
15. Childs GV, Lloyd JM, Unabia G, Charib SD, Wierman ME, Chin WW (1987) Detection of luteinizing hormone B messenger ribonucleic acid (RNA) in individual gonadotropes after castration: Use of a new in situ hybridization method with a photobiotinylated complementary RNA probe. Mol Endocrinol 1:926–932
16. Cline MJ, Battifora H (1987) Abnormalities of protooncogenes in non-small cell lung cancer. Correlations with tumor type and clinical characteristics. Cancer 60:2669–2674
17. Coghlan JP, Aldred P, Haralambidis J, Niall HD, Penschow JD, Tregear GW (1985) Hybridization histochemistry. Anal Biochm 149:1–28
18. Cohen PS, Seeger RC, Triche TJ, Israel MA (1988) Detection of N-*myc* gene expression in neuroblastoma tumors by *in situ* hybridization. Am J Pathol 131:391–397
19. Cox KH, DeLeon DV, Angerer LM, Angerer RC (1984) Detection of mRNAs in sea urchin embryos by in situ hybridization using asymmetric RNA probes. Dev Biol 101:485–502

19a. Daughaday WH, Emanuele MA, Brooks MH, Barbato AL, Kapadia M, Rotwein P (1988) Synthesis and secretion of insulin-like growth factor II by a leiomyosarcoma with associated hypoglycemia. N Engl J Med 319:1434–1440

20. De Keyzer Y, Bertagna X, Lenne F, Girard F, Luton J, Kahn A (1985) Altered proopiomelanocortin gene expression in adrenocorticotropin-producing nonpituitary tumors: comparative studies with corticotropic adenomas and normal pituitaries. J Clin Invest 76:1892–1898
21. DeLellis RA, Wolfe HJ (1987) New techniques in gene product analysis. Arch Pathol Lab Med 111:620–627
22. Fagin JA, Brown A, Melmed S (1988) Regulation of pituitary insulin-like growth factor I messenger ribonucleic acid levels in rats harboring somatomammotropic tumors: implications for growth hormone autoregulation. Endocrinology 122: 2204–2210
23. Gall JG, Pardue ML (1969) Formation and detection of RNA-DNA hybrid molecules in cytological preparation. Proc Natl Acad Sci USA 63:378–383
24. Gee CE, Chen CL, Roberts JL, Thompson R, Watson SJ (1983) Identification of proopiomelanocortin neurones in rat hypothalamus by *in situ* cDNA-mRNA hybridization. Nature 306:374–376
25. Gee CE, Roberts JL (1983) In situ hybridization histochemistry: A technique for the study of gene expression in single cells. DNA 2:157–163
26. Hagn C, Schmid KW, Fischer-Colbrie R, Winkler H (1986) Chromogranin A, B and C in human adrenal medulla and endocrine tissues. Lab Invest 55: 405–411
27. Hamid QA, Bishop AE, Sikki KL, Varndell IM, Bloom SR, Polak JM (1986) Immunocytochemical characterization of 10 pancreatic tumors associated with the glucagonoma syndrome, using antibodies to separate regions of the pro-glucagon molecule and other neuroendocrine markers. Histopathology 10:119–133
28. Hamid QA, Bishop AE, Springall DR, Adams C, Giaid A, Denny P, Ghatei M, Legon S, Cuttita F, Rode J, Spindel E, Bloom SR, Polak JM (1989) Detection of human probombesin mRNA in neuroendocrine (small cell) carcinoma of the lung. *In situ* hybridization with cRNA probe. Cancer 63:266–271
29. Hankin RC, Lloyd RV (1989) Detection of messenger RNA in routinely processed tissue sections with biotinylated oligonucleotide probes. Am J Clin Pathol 92:166–171
30. Harrison PR, Conkie D, Paul J (1973) Localization of cellular globin messenger RNA by *in situ* hybridization to complementary DNA. FEBS Lett 32: 109–115
31. Haselbacher GK, Irminger J-C, Zapf J, Ziegler WH, Humbel RE (1987) Insulin-like growth factor II in human adrenal pheochromocytomas and Wilms tumors: Expression at the mRNA and protein level. Proc Natl Acad Sci USA 84:1104–1106
32. Helman LJ, Gazdar AF, Park J-G, Cohen PS, Cotelingam JD, Israel MA (1988) Chromogranin A expression in normal and malignant human tissues. J Clin Invest 82:686–690
33. Hofler H, Childers H, Lechan RM, Montminy MR, Goodman RH, Wolfe HJ (1989) Distribution of somatostatin mRNA containing neurons in the rat hypothalamus demonstrated by in situ hybridization with antisense RNA probes. J Histochem Cytochem (in press)
34. Hofler H, Childers H, Montminy MR, Lechan RM, Goodman RH, Wolfe HJ (1986) *In situ*

hybridization methods for the detection of somatostatin mRNA in tissue sections using antisense RNA probes. Histochem J 18:597–604

35. Hofler H, Klimpfinger M, Ruhrich P, Pfranger R, Spindel ER, Putz B, Wirnsberger G (1987) Medullary thyroid carcinoma: characterization of neuroendocrine gene and oncogene expression. Lab Invest 56:31A
36. Hofler H, Putz B, Ruhn C, Wirmsberger G, Klimptinger M, Smalle J (1987) Simultaneous localization of calcitonin mRNA and peptide in a medullary thyroid carcinoma. Virchows Arch [Cell Pathol] 54: 144–151
37. Hoppener JWM, Steenbergh PH, Moonen PJJ, Wagenaar SS, Jansz HS, Lips CJM (1986) Detection of mRNA encoding calcitonin, calcitonin gene related peptide and proopiomelanocortin in human tumors. Mol Cell Endocrinol 47:125–130
38. Hoppener JWM, Steenbergh PH, Slebos RJC, Visser A, Lips CJM, Jansz HS, et al. (1987) Expression of the second calcitonin/calcitonin gene-related peptide gene in Ewing's sarcoma cell lines. J Clin Endocrinol Metab 64:809–817
39. Hudson P, Penschow J, Shine J, Ryan G, Niall H, Coghlan J (1981) Hybridization histochemistry: Use of recombinant DNA as a "homing probe" for tissue localization of specific mRNA populations. Endocrinology 108:353–356
40. Iacangelo A, Affolter HU, Eiden LE, Herbert E, Grimes M (1986) Bovine chromogranin A sequence and distribution of its messenger RNA in endocrine tissues. Nature 323:82–86
41. Iacangelo A, Fischer-Colbrie R, Koller KJ, Brownstein MJ, Eiden LE (1988) The sequence of porcine chromogranin A messenger RNA demonstrates chromogranin A can serve as the precursor for the biologically active hormone, pancreastatin. Endocrinology 122:2339–2341
42. Iacangelo A, Okayama H, Eiden LE (1988) Primary structure of rat chromogranin A and distribution of its mRNA. FEBS Lett 227:115–121
43. Ikeda K, Weir EC, Mangin M, Dannies PS, Kinder B, Deftos LJ, Brown EM, Broadus AE (1988) Expression of messenger ribonucleic acid encoding a parathyroid hormone-like peptide in normal human and animal tissues with abnormal expression in human parathyroid adenoma. Mol Endocrinol 2:1230–1236
44. Jameson JL, Klibanski A, Black PM, Zervas NT, Lindell CM, Hsu DW, Ridgeway EC, Habener JF (1987) Glycoprotein hormone genes are expressed in clinically nonfunctioning pituitary adenomas. J Clin Invest 80:1472–1478
45. Jameson JL, Lindell CM, Habener JF (1987) Gonadotropin and thyrotropin α and β-subunit gene expression in normal and neoplastic tissues characterized using specific messenger ribonucleic acid hybridization probes. J Clin Endocrinol Metab 64: 319–327
46. Jeffrey W (1982) Messenger RNA in the cytoskeletal framework. Analysis by in situ hybridization. J Cell Biol 95:1–7
47. Jirikowski GF, Ramalho-Ortigao JF, Lindl T, Seliger H (1989) Immunocytochemistry of 5-bromo-2-deoxyuridine labelled oligonucleotide probes. A novel technique for *in situ* hybridization. Histochemistry 91:51–53
48. John HA, Patrinou-Georgoulas M, Jones KW (1981) Detection of myosin heavy chain mRNA during myogenesis in tissue culture by in vitro and in situ hybridization. Cell 21:501–508
49. Kamata T, Feramisco JR (1984) Epidermal growth factor stimulates genuine nucleotide binding activity and phosphorylation of *ras* oncogene proteins. Nature 310:147–150
50. Kovacs K, Lloyd RV, Horvath E, Asa SL, Stefaneanu L, Killinger DW, Smythe HS (1989) Silent somatotroph adenomas of the human pituitary. A morphologic study of cases including immunocytochemistry, electron microscopy and *in vitro* examination and *in situ* hybridization. Am J Pathol 134: 345–353
51. Lawrence JB, Singer RH (1985) Quantitative analysis of *in situ* hybridization methods for the detection of actin gene expression. Nucleic Acid Res 13:1777–1799
52. Lemoine NR, Mayall ES, Wyllre FS, Farr CJ, Hughes D, Padua RA, Thurston V, Williams ED, Wynford-Thomas D (1988) Activated *ras* oncogenes in human thyroid cancers. Cancer Res 48: 4459–4463
53. Lewis ME, Sherman TG, Watson SJ (1987) In situ hybridization histochemistry with synthetic oligonucleotides: strategies and methods. Peptides 6 (Suppl 2):75–87
54. Lloyd RV (1987) Use of molecular probes in the study of endocrine diseases. Hum Pathol 18:1199–1211
55. Lloyd RV (1988) Analysis of human pituitary tumors by *in situ* hybridization. Pathol Res Pract 183:558–560
56. Lloyd RV, Cano M, Chandler WF, Barkan AL, Horvath E, Kovacs K (1989) Human growth hormone and prolactin secreting pituitary adenomas analyzed by in situ hybridization. Am J Pathol 134:605–613
57. Lloyd RV, Iacangelo A, Eiden LE, Cano M, Jin L, Grimes M (1989) Chromogranin A and B mes-

senger ribonucleic acids in pituitary and other normal and neoplastic human endocrine tissues. Lab Invest 60:548–556

58. Lloyd RV, Landefeld TD (1986) Detection of prolactin messenger RNA in rat anterior pituitary by *in situ* hybridization. Am J Pathol 125:35–44
59. Lloyd RV, Mervak T, Schmidt K, Warner TFCS, Wilson BS (1984) Immunohistochemical detection of chromogranin and neuron-specific enolase in pancreatic endocrine neoplasms. Am J Surg Pathol 8:607–614
60. Lloyd RV, Wilson BS (1983) Specific endocrine marker defined by a monoclonal antibody. Science 222:628–630
61. Melmed S, Ezrin C, Kovacs K, Goodman RS, Frohman LA (1983) Acromegaly due to secretion of growth hormone by an ectopic pancreatic islet cell tumor. N Engl J Med 312:9–16
62. Melmed S, Yamashita S, Kovacs K, Ong J, Rosenblah S, Brannstein G (1987) Cushing's syndrome due to ectopic pro-opiomelanocortin gene expression by islet cell carcinoma of the pancreas. Cancer 59:772–778
63. Morel G, Dubois P, Gossard F (1986) Détection ultrastructurale des ARN messagers codant pour l'hormone de croissance dans l'antéhypophyse du rat par hybridation in situ. CR Acad Sci Paris 302 (III):479–484
64. Morely DJ, Hodes ME (1987) In situ localization of amylase mRNA and protein. An investigation of amylase gene activity in normal human parotid gland. J Histochem Cytochem 35:9–14
65. Morimoto H, Monden T, Shimano T, Higashiyama M, Tomita N, Murotani M, Matsuura N, Okuda H, Mori T (1987) Use of sulfonated probes for *in situ* detection of amylase mRNA in formalin-fixed paraffin sections of human pancreas and submaxillary gland. Lab Invest 57:737–741
66. Murphy LJ, Bell GI, Friesen HG (1987) Growth hormone stimulates sequential induction of *c-myc* and insulin-like growth factor I expression in vivo. Endocrinology 120:1806–1812
67. Nakanishi S, Inoue A, Kita T, Nakamura M, Chang ACY, Cohen SN, Numa S (1979) Nucleotide sequence of cloned cDNA for bovine corticotropin-B-lipotropin precursor. Nature 278:423–427
68. Nogami H, Yoshimura F, Carrillo AJ, Sharp ZD, Sheridan PJ (1985) Estrogen induced prolactin mRNA accumulation in adult rat pituitary as revealed by *in situ* hybridization. Endocrinol Jpn 32: 625–634
69. O'Connor DT, Deftos LJ (1986) Secretion of chromogranin A by peptide-producing endocrine neoplasms. N Engl J Med 314:1145–1151
70. Pintar JE, Schachter BS, Herman AB, Durserian S, Krieger DT (1984) Characterization and localization of pro-opiomelanocortin messenger RNA in the adult rat testes. Science 225:632–634
71. Pixley S, Weiss M, Melmed S (1987) Identification of human growth hormone messenger ribonucleic acid in pituitary adenoma cells by in situ hybridization. J Clin Endocrinol Metab 65:575–580
72. Pochet R, Brocas H, Vassart G, Toubeau G, Seyo H, Refetoff S, Dumont JE, Pasteels JL (1981) Radioautographic localization of prolactin messenger RNA on histological sections by *in situ* hybridization. Brain Res 211:433–438
73. Priestley JV, Hynes MA, Han VKM, Rethely M, Perl ER, Lund PK (1988) In situ hybridization using ^{32}P labelled oligodeoxyribonucleotides for the cellular localisation of mRNA in neuronal and endocrine tissue. An analysis of procedure variables. Histochemistry 89:467–479

73a. Sakunari T, Seo H, Yamamoto N, Nagaya T, Nakane T, Kuwayama A, Kageyama N, Matsui N (1988) Detection of mRNA of prolactin and ACTH in clinically nonfunctioning pituitary adenomas. J Neursurg 69:653–659

74. Salomon DS, Perroteau I (1986) Growth factors in cancer and their relationship to oncogenes. Cancer Invest 4:43–60
75. Sherman TG, Civelli O, Douglass J, Herbert E, Watson SJ (1986) Coordinate expression of hypothalamic pro-dynorphin and pro-vasopressin mRNAs with osmotic stimulation. Neuroendocrinology 44: 222–228
76. Shivers BD, Harlan RE, Pfaff DW, Schachter BS (1986) Combination of immunocytochemistry and in situ hybridization in the same tissue section of rat pituitary. J Histochem Cytochem 34:39–43
77. Simpson EL, Mundy GR, D'Souza SM, Ibbotson KJ, Bockman R, Jacobs JW (1983) Absence of parathyroid hormone messenger RNA in nonparathyroid tumors associated with hypercalcemia. N Engl J Med 309:325–330
78. Singer RH, Ward DC (1982) Actin gene expression visualized in a chicken muscle tissue culture by using in situ hybridization with a biotinated nucleotide analog. Proc Natl Acad Sci USA 79:7331–7335
79. Southern EM (1975) Detection of specific sequences among DNA fragments separated by gel electrophoresis. J Mol Biol 98:503–517
80. Steenbergh PH, Hoppener JWM, Zandberg J, Roos BA, Jansz HS, Lips JM (1984) Expression of the proopiomelanocortin gene in human medullary thyroid carcinoma. J Clin Endocrinol Metab 58: 904–908

80a. Stork PJ, Herteaux C, Frazier R, Kronenburg H, Wolfe HJ (1989) Expression and distribution of parathyroid hormone messenger RNA in pathological conditions of the parathyroid. Lab Invest 60:92A (abstract 547)
81. Stornetta RL, Hawelu-Johnson CL, Guyenet PG, Lynch KR (1988) Astrocytes synthesize angiotensinogen in brain. Science 242:1444–1446
82. Suva LJ, Winslow GA, Wettenhall EH, Hammonds RG, Moseley JM, Dietenbach-Jagger H, Rodda CP, Kemp BE, Rodriguez H, Chen EY, Hudson PJ, Martin TJ, Wood WI (1987) A parathyroid hormone-related protein implicated in malignant hypercalcemia: Cloning and expression. Science 237: 893–896
83. Thomas PS (1980) Hybridization of denatured RNA and small DNA fragments transferred to nitrocellulose. Proc Natl Acad Sci USA 77:5201–5205
84. Uhl GR, Reppert SM (1986) Suprachiasmatic nucleus vasopressin messenger RNA—circadial variation in normal and Brattleboro rats. Science 232:390–393
85. Uhl GR, Zingg HH, Habener JF (1985) Vasopressin mRNA *in situ* hybridization: localization regulation studied with oligonucleotide cDNA probes in normal and Brattleboro rat hypothalamus. Proc Natl Acad Sci USA 82:5555–5559
86. Varndell IM, Polak JM, Sikri KL, Minth CD, Bloom SR, Dixon JE (1984) Visualization of messenger RNA directing peptide synthesis by in situ hybridization using a novel single-stranded cDNA probe. Potential for the investigation of gene expression and endocrine cell activity. Histochemistry 81:597–601
87. Wilcox JN, Gee CE, Roberts JL (1986) In situ cDNA: mRNA hybridization: Development of a technique to measure mRNA levels in individual cells. Methods Enzymol 124:510–533
88. Wolfson B, Manning RW, Davis LG, Arentzen R, Baldino F Jr (1986) Co-localization of corticotropin releasing factor and vasopressin mRNA in neurones after adrenalectomy. Nature 315: 59–61
89. Yamamoto N, Seo H, Suganuma N, Matsui N, Nakane T, Kuwayama A, Kageyama N (1986) Effect of estrogen on prolactin mRNA in the rat pituitary. Analysis by in situ hybridization and immunohistochemistry. Neuroendocrinology 42: 494–497
90. Zabel M, Schafer H (1988) Localization of calcitonin and calcitnonin gene-related peptide mRNAs in rat parafollicular cells by hybridocytochemistry. J Histochem Cytochem 36:543–546
91. Zajac JD, Penschow J, Mason T, Tregear G, Coghlan J, Martin JJ (1986) Identification of calcitonin in calcitonin gene-related peptide messenger ribonucleic acid in medullary thyroid carcinomas by hybridization histochemistry. J Clin Endocrinol Metab 62:1037–1043

Hybridization Protocols

DNA-RNA Oligonucleotide for Frozen Sections

Tissues are cut at 6 μm, fixed in 3% paraformaldehyde for 20 minutes, dehydrated, then stored at 70°C.

Day 1 (Use RNase-free conditions):

1. 100% ethanol (EtOH) 5 min.
2. 95% EtOH 5 min.
3. 2 × SSC 3 min (1 × SSC = 0.15 M NaCl and 0.015 M Na citrate).
4. Acetic anhydride + triethanolamine: 625 μl acetic anhydride for 10 min, 250 ml triethanolamine.
5. 2 × SSC, 70°C, 30 min.
6. 2 × SSC (room temperature), 3 min.
7. Hybridization mixture 40μl, 1 hour.
8. Add 100,000 to 1,000,000 cpm ^{35}S or ^{3}H in hybridization medium with dithiothreitol (3 mg/10 ml).
9. Put on RNase-free coverslips and incubate at 44°C for 12 to 18 hours.

Day 2:

10. Remove coverslips in 2 × SSC (does *not* have to be RNase-free). Shake slides while washing.
11.* Wash at room temperature, 2 × SSC, 2 hours. Change 2 × SSC after 5 min twice, then after 30 min until 2 hours of washing is completed.
12. Wash at room temperature, 1 × SSC, 1 hour. Change after 30 min.
13. Wash at room temperature, 0.5 × SSC, 30 min, then 30 min at 44°C.

---

*If immunoperoxidase is to be used, start this procedure after 1 hour in 2 × SSC. Other slides from the same experiment should remain in 2 × SSC for an additional hour during the immunoperoxidase staining and should then be washed in 1 × SSC. Note that with the immunoperoxidase staining after the DAB step the slides should be rinsed in 2 × SSC, *not* H_2O, and the washing procedure for hybridization continued from step 13 to step 16.

14. Wash at room temperature, 0.5 × SSC, 30 min.
15. Alcohol 70% for 2 min with 0.3 M NH_4 acetate, 100% for 2 min with 0.3 M NH_4 acetate.
16. Air-dry, then autoradiography.

DNA-RNA Oligonucleotide for Paraffin Sections

Formalin-fixed, paraffin-embedded blocks are cut at 3 to 4 μm.

Day 1 (Use RNase-free conditions):

1. Xylene 10 min (this step is different from the protocol for frozen sections).
2. 100% EtOH 5 min.
3. 95% EtOH 5 min.
4. 2 × SSC for 3 min (1 × SSC = 0.15 M NaCl and 0.015 M Na citrate).
5. 0.2 N HCl 20 min (this step different from protocol for frozen sections).
6. 2 × SSC (two times) 3 min each.
7. Acetic anhydride + triethanolamine for 10 min: 625 μl acetic anhydride for 10 min, 250 ml triethanolamine.
8. 2 × SSC, 70°C, 30 min.
9. 2 × SSC (room temperature) 3 min.
10. Hybridization mixture, 40 μl, 1 hour.
11. Add 100,000 to 2,000,000 cpm ^{35}S or ^{3}H in hybridization medium with dithiothreitol (3 mg/10 ml).
12. Put on RNase-free coverslips and incubate at 44°C for 12 to 18 hours.

Day 2:

13. Remove coverslips in 2 × SSC (does *not* have to be sterile). Shake slides while washing.
14.* Wash at room temperature, 2 × SSC, 2 hours. Change 2 × SSC after 5 min twice, then after 30 min until washing for 2 hours is completed.
15. Wash at room temperature, 1 × SSC, 1 hour. Change after 30 min.
16. Wash at room temperature, 0.5 × SSC, 30 min.
17. Wash at room temperature, 0.5 × SSC, 30 min.
18. Alcohol 70% for 2 min with 0.3 M NH_4 acetate, 100% for 2 min with 0.3 M NH_4 acetate.
19. Air-dry, then autoradiography.

*If immunoperoxidase is to be used, start this procedure after 1 hour in 2 × SSC. Other slides from the same experiment should remain in 2 × SSC for an additional hour during the immunoperoxidase staining and should then be washed in 1 × SSC. Note that with the immunoperoxidase staining after the DAB step the slides should be rinsed in 2 × SSC, *not* H_2O, and the washing procedure for hybridization continued from step 16 to step 19.

Combined In Situ-Immunoperoxidase Procedures

Perform Day 1 procedure as above. On Day 2, after washing for 1 hour in 2 × SSC, start immunoperoxidase procedure.

1. Treat with H_2O_2 material (1 : 1) for 5 min.
2. Wash in 2 × SSC for 3 min.
3. Suppressor serum, 10 min.
4. Shake off suppressor serum. Add primary antibody. Incubate for 30 min at appropriate dilution.
5. Wash twice, 2 min each, in 2 × SSC.
6. Biotin IgG, 30 min.
7. Wash twice, 2 min each, in 2 × SSC.
8. ABC complex, 30 min.
9. Wash twice, 2 min each, in 2 × SSC.
10. DAB, 5 min (cytospins) or 10 min (tissues): 10 mg DAB/25 ml PBS.
11. Wash in 2 × SSC for 2 min.
12. Wash in 1 × SSC for 2 min.
13. Go to step 13 (frozens) or 16 (paraffins) of Day 2 in situ procedure and continue from there.

In Situ Hybridization with Antisense RNA Probe

Day 1 (Use RNase-free conditions):

1. 100% EtOH, 2 min.
2. 95% EtOH, 2 min.
3. 2 × SSC, 5 min.
4. 0.3% Triton X-100 (in PBS), 15 min.
5. 2 × SSC, 2 min.
6. Proteinase K, 37°C, 20 min.
7. 4% paraformaldehyde, 5 min.
8. 2 × SSC, 2 min.
9. Acetic anhydride + triethanolomine, 10 min: 625 μl acetic anhydride, 250 ml triethanolamine.
10. 2 × SSC, 2 min.

APPENDIX III
Weights of Some Endocrine Organs

Tissue/diagnosis	Weight range
Adrenal (adults)	
Adrenal gland (4)	3.5–4.5 g
Adrenal medulla (1,6)	0.4–0.5 g
Adrenal in Cushing's–pituitary tumor (5)	6–12 g
Adrenal in Cushing's secondary to ectopic ACTH (5)	12–20 g
Pituitary	
Neonate	100 mg
Children (10–20 years) (7)	Mean of 560 mg
Adult male	350–600 mg
Adult female nulliparous	400–900 mg
Adult female pregnant	840–1060 mg
Parathyroid	
Neonates (each gland)	4–20 mg
Adults (four glands) (7)	120–180 mg
Thyroid (2,3)	
Neonate	1.5 g
Adult	14–20 g

References

1. DeLellis RA, Wolfe HJ, Gagel RF, Feldman BS, Miller HH, Gang DL, Reichlin S (1976) Adrenal medullary hyperplasia. A morphometric analysis in patients with familial medullary thyroid carcinoma. Am J Pathol 83:177–196
2. Kay C, Abrahams S, McClain P (1966) The weight of normal thyroid glands in children. Arch Pathol 82:349–352
3. Mochizuki Y, Mowafy R, Pasternack B (1963) Weights of human thyroids in New York City. Health Phys 9:1299–1301
4. Nichols J (1968) Adrenal cortex. In: Bloodworth JMB Jr (ed) Endocrine Pathology, General and Surgical. Baltimore, Williams & Wilkins, pp 224–255
5. Page DL, De Lellis RA, Hough AJ Jr (1986) Tumors of the adrenal. Atlas of Tumor Pathology, Second Series, Fascicle 23. Washington DC, Armed Forces Institute of Pathology
6. Quinan C, Berger AA (1933) Observations on human adrenals with especial reference to the relative weight of the normal medulla. Ann Intern Med 6:1180–1192
7. Sunderman FW, Boerner F (1949) Normal Values in Clinical Medicine. Philadelphia, Saunders

Index

T

Z